Visualisation and Epistemological Access to Mathematics Education in Southern Africa

This book demonstrates that using visualisation processes in mathematics education can help to enhance teaching and learning and bridge the inequality gap that exists between well-resourced and under-resourced schools in Southern Africa.

Drawing on classroom research conducted in the Southern African region, it examines how epistemological access in a context of gross inequality can be constructively addressed by providing research-based solutions and recommendations. The book outlines the visualisation process as an integral but often overlooked process of mathematics teaching and learning. It goes beyond the traditional understanding of visualisation processes such as picture forming and using tools and considers visualisation processes that are semiotic in nature and includes actions such as gestures in combination with language. It adds value to the visualisation in mathematics education research discourse and deliberation in Africa.

With a unique focus on Southern Africa and open avenues for further research and collaboration in the region, it will be a highly relevant reading for researchers, academics and post-graduate students of mathematics education, comparative education and social justice education.

Marc Schäfer was formally Professor of Mathematics Education and SARChI Chair in Mathematics Education, Rhodes University, South Africa.

Routledge Research in STEM Education

The *Routledge Research in STEM Education* series is home to cutting-edge, upper-level scholarly studies and edited collections covering STEM Education.

Considering science, technology, engineering, and mathematics, texts address a broad range of topics including pedagogy, curriculum, policy, teacher education, and the promotion of diversity within STEM programmes.

Titles offer dynamic interventions into established subjects and innovative studies on emerging topics.

Science and Technology Teacher Education in the Anthropocene
Addressing Challenges in the South and North
Edited by Miranda Rocksén, Elaosi Vhurumuku, Maria Svensson, Emmanuel Mushayikwa and Audrey Msimanga

Perspectives in Contemporary STEM Education Research
Research Methodology and Design
Edited by Thomas Delahunty and Máire Ní Ríordáin

Teaching Assistants, Inclusion and Special Educational Needs:
International Perspectives on the Role of Paraprofessionals in Schools
Edited by Rob Webster & Anke A. de Boer

Invention Pedagogy – The Finnish Approach to Maker Education
Edited by Tiina Korhonen, Kaiju Kangas and Laura Salo

Visualisation and Epistemological Access to Mathematics Education in Southern Africa
Edited by Marc Schäfer

For more information about this series, please visit: www.routledge.com/Routledge-Research-in-STEM-Education/book-series/RRSTEM

Visualisation and Epistemological Access to Mathematics Education in Southern Africa

Edited by
Marc Schäfer

LONDON AND NEW YORK

First published 2023
by Routledge
4 Park Square, Milton Park, Abingdon, Oxon OX14 4RN

and by Routledge
605 Third Avenue, New York, NY 10158

Routledge is an imprint of the Taylor & Francis Group, an informa business

© 2023 selection and editorial matter, Marc Schäfer; individual chapters, the contributors

The right of Marc Schäfer to be identified as the author of the editorial material, and of the authors for their individual chapters, has been asserted in accordance with sections 77 and 78 of the Copyright, Designs and Patents Act 1988.

All rights reserved. No part of this book may be reprinted or reproduced or utilised in any form or by any electronic, mechanical, or other means, now known or hereafter invented, including photocopying and recording, or in any information storage or retrieval system, without permission in writing from the publishers.

Trademark notice: Product or corporate names may be trademarks or registered trademarks, and are used only for identification and explanation without intent to infringe.

British Library Cataloguing-in-Publication Data
A catalogue record for this book is available from the British Library

ISBN: 978-1-032-00040-4 (hbk)
ISBN: 978-1-032-00042-8 (pbk)
ISBN: 978-1-003-17242-0 (ebk)

DOI: 10.4324/9781003172420

Typeset in Galliard
by codeMantra

Dedicated to Oscar
and also to Finley, Sophie, Remy, Tiger

Contents

Acknowledgements

I wish to record my thanks to and acknowledgment of the following people who made invaluable contributions to this book:

- all the authors who were willing to write their stories and share their research experiences;
- all the participants of the research projects that gave substance to the visualisation theme of this book;
- Jean Schäfer, for her meticulous proofreading and willingness in preparing the manuscript for Routledge;
- Jaine Roberts, the Director of the Research Office at Rhodes University, for her institutional support;
- all the Routledge staff for their support and advice;
- the South African National Research Foundation (NRF) in providing significant financial support for the research activities of many of the researchers in this book. The opinions and arguments of the authors are not necessarily those of the NRF.

Editors and Contributors

Charity Ausiku is a lecturer and former Teaching Practice (TP) Coordinator at the University of Namibia, Rundu Campus. She also served as a Research Coordinator for final year students enrolled for the Bachelor of Education (BEd) degree. Having completed her HED (secondary) at the University of Namibia, she acquired her BEd (Honours) and MEd in Mathematics Education from Rhodes University. Her main research interest lies in the field of Mathematics Education and Teaching Practice. Her keen interest in visualisation as an alternative approach to teaching mathematics emanates from her wide experience in the teaching fraternity.

Clemence Chikiwa is currently a lecturer at Sol Plaatjie University. He is a PhD graduate from Rhodes University where he worked as a Senior Lecturer and Researcher in the Education Department and the SARChI Mathematics Education Chair. During the five years he worked in the Chair, a strong visualisation focus developed, and a number of master's degree students researched and graduated under his supervision. Clemence has done both under-graduate and post-graduate teaching and supervision in South Africa, Namibia and Zimbabwe. Having learnt and studied in a second language from primary school, his main research interests are in mathematics education and language-related aspects. He is also interested in the use of visualisation processes and technological tools in mathematics education and their relationship to language issues. Clemence believes that language, if appropriately used in mathematics, can be a powerful visualisation tool to enhance conceptual understanding.

Beata Dongwi taught high school mathematics passionately for many years and produced excellent results in national examinations. She obtained her PhD in mathematics education from Rhodes University. Her research interests include visualisation in mathematics classrooms, particularly in how students use their bodies to enact meaning. She is keenly interested in how students incorporate their holistic being in problem solving and in ways that best interpret various visual representations that students bring forth in mathematical activities. She served as an executive member of the Southern African Association for Research in Mathematics, Science and Technology Education. She hopes to inspire more novice researchers and mathematicians to pursue their goals and contribute to the education landscape. She is currently a Bachelor of Science in Applied Mathematics and Statistics student at the Namibia University of Science and Technology. In her spare time, she tutors in-service mathematics teachers at the University of Namibia and assesses their mathematics education and research assignments.

Ronald Griqua is Project Facilitator of the Rhodes University Mathematics Education Project (RUMEP) at Rhodes University, Grahamstown/Makhanda, South Africa. He was also an educator at the Department of Education for most of his working life. He was furthermore employed as a Human Resources Development Specialist at BHPBilliton and Kumba Iron Ore. He graduated from the University of the Western Cape and has post-graduate degrees from the University of the Western Cape and Rhodes University. His passion for teaching is fully lived out in the role that he plays in RUMEP, as it deals with in-service teacher development. The engagement with RUMEP ignited his research interest in visualisation in mathematics education.

Sindisiwe Herbert is a mathematics teacher in Cape Town, South Africa. She holds bachelors and Honours degrees from Rhodes University and is completing a master's degree in mathematics education. Her passion for mathematics and teaching were ignited when she read Fermat's Last Theorem and was suddenly aware of the breadth and beauty of a field that was so much deeper than practising rote algorithms and learning dry theorems. Since then her focus has turned to visualisation as a means of helping students see and appreciate the deeper structures behind the mathematics that they are forced to learn at school, particularly algebra.

Matthias Ludwig holds a master's degree in teacher education for mathematics and physics for lower and upper secondary school. During his time as a teacher he finished his PhD dissertation at the University of Würzburg. From 2002 until 2011 he was a professor at the University of Education in Weingarten. Since March 2011 he has been a full professor for mathematics education at the Goethe University in Frankfurt and is currently Head of the Institute for Mathematics Education and Computer Sciences Education. In the German Society of Mathematics Education he once led a geometry group, and was responsible for organising their annual conference. His research interests lie in learning space geometry and the usage of pictures, models and computer animation by solving space geometry tasks. Currently he coordinates different European projects about mobile learning in mathematics education (www.momatre.eu, www.masce.eu, www.colette-project.eu and www.asymptote-project.eu) with ten different European countries as partners. In all these projects, where the mobile device is the link between the learner and the reality, visualisation in mathematics plays a crucial role.

Deepak Mavani is a freelance researcher in mathematics education based in Thiruvananthapuram, India. He was Head of the Department of Mathematics at an NGO school in Mthatha, Eastern Cape, South Africa. He recently graduated with his doctoral degree from Rhodes University, South Africa. Earlier he graduated with a post-graduate degree from the University of Kerala, India. He initiated a teacher development project, GLIP, in Mthatha, for integrating technological tools in classrooms. He enjoys and values the use of visualisation of mathematical concepts and problem solving. He argues that as students gain familiarity with visual techniques, they gain confidence, excelling in mathematics, and start to appreciate mathematics. He sees technology-integrated visualisation in particular as an ideal opportunity to provide 'mathematics for all'.

Vimolan Mudaly is a Professor of Mathematics Education at the University of KwaZulu-Natal. He has had an illustrious career in education, at both secondary school and tertiary levels. Besides teaching in the under- and post-graduate sectors, he has graduated a multitude of master's and PhD students. Vimolan has published widely and was rewarded for being among the top 30 researchers in his institution. Recently, he was selected to compile the Covid-19 report for education in South Africa. His research began with mathematical modelling and progressed naturally to visualisation. Currently, much of his research focuses on visual pedagogies that are aimed at improving teacher effectiveness in mathematics classrooms.

Cristina Sabena is Associate Professor at the Department of Philosophy and Science Education of the University of Torino, where she teaches Mathematics Education to future primary school teachers. She is currently the Secretary of CIEAEM (The International Commission for the Study and Improvement of Mathematics Teaching), and member of CIIM (a scientific commission of Italian Mathematics Association, dealing with Mathematics teaching). She obtained her PhD in mathematics education at the University of Torino, with a thesis on developing a multimodal semiotic approach to the teaching-learning of calculus. Her interest in visualisation stems from these early studies on the role of embodied resources as signs in mathematics. She also focuses her research on the networking of theories in mathematics education, and, more generally, on the role of theoretical approaches in the field and on the development of argumentative skills as a means of critical citizenship.

She has collaborated with numerous colleagues in Canada, France, Germany, Israel and South Africa. She also contributes to teacher development projects in Italy. Recently she engaged in conjugating informal mathematics education contexts within a teacher professional project.

Duncan Samson is Director of Mathematics at St Andrew's College and the Diocesan School for Girls in Makhanda (Grahamstown), South Africa. He is editor of AMESA's teacher practitioner journal *Learning and Teaching Mathematics* and is coordinator of the Junior Round 1 & 2 Committee of the South African Mathematics Olympiad. He holds a PhD in Mathematics Education from Rhodes University where his thesis explored the heuristic significance of enacted visualisation. He was previously a researcher for the FRF Mathematics Education Chair at Rhodes University where he played a pivotal role in the Chair's Mathematics Teacher Enrichment Programme. He was also actively involved in the VITALmaths project – a multilingual collaborative research and development project between the University of Applied Sciences Northwestern Switzerland (FHNW) and Rhodes University. Duncan is actively involved in resource development, and continues to publish regularly in the field of mathematics education. His passion for visualisation in mathematics stems from a sensitivity to a multi-representational view of the subject as well as a desire to gain a deeper insight into pupils' mathematical sense-making.

Marc Schäfer is Professor and former Head of Department of Education at Rhodes University, Grahamstown/Makhanda, South Africa. He was also National Secretary of the Association of Mathematics Educators of South Africa (AMESA), and the National President of the Southern African Association for Research in Mathematics, Science and Technology Education (SAARMSTE). He holds the SARChI Chair for Mathematics Education at Rhodes University and has published widely, and has supervised over 80 master's and PhD students to completion. He graduated from the University of the Witwatersrand and has post-graduate degrees from Rhodes University and a PhD from Curtin University in Perth. He was recently a visiting professor at the Goethe University in Germany and has collaborated in the field of Mathematics Education with researchers in Switzerland, Germany, Namibia, India, Italy and Canada. He has also coordinated many teacher development projects in South Africa and Namibia. His main research interest in visualisation in mathematics stems from his passion about reforming mathematics pedagogy to make mathematics more visible not only within itself, but also within the classroom culture.

Visualisation and Epistemological Access to Mathematics Education in Southern Africa

Ferdinando Arzarello
Dept. of Mathematics
Turin University

I have been impressed by this book because of the complexity of the issues it affords and intertwines: in fact, it is a book of mathematics education but, as sometimes happens within this context, it also raises important socio-political problems. These are, of course, interwoven with the specificity of South African society and with endeavours for improving the school system within a democratic and equitable stance after the apartheid period, but it also has the advantage of merging the discussion into a broader international context, touching upon issues that affect all countries in the world.

As I learnt from some of the chapters, the book gives substance in mathematical education to the reflections of scholars like Wally Morrow, who from the seventies was a vigorous and independent voice on educational policy and practice in South Africa, and in his work pointed out that the culture of *entitlement* in education can be incoherent with the concept of educational *achievement.* In 2002 he coined the term 'epistemological access', a construct which since then has become a must in the scholarship of higher education learning in South Africa. As written by J. Muller (2014, p. 255), "the concept of epistemological access has since been deployed in numerous papers and publications as a banner to signal intent to move beyond physical or formal access to meaningful access to the 'goods' of the university". The main task for schools is that all people can gain this type of access, that is, "learning how to become a participant in a practice, and since academic practices have developed around the search for knowledge, access to an academic practice entailed epistemological access. To deny this was 'facile relativism'" (Morrow, 2009, p. 70). This means focusing both on *know how* (procedural) knowledge and on *know that* (propositional) knowledge, that is, *doing things with knowledge.*

Such an epistemological, cognitive and pedagogical framework is at the heart of the proposals that constitute the 11 chapters of this book. Specifically, visualisation constitutes the powerful tool through which mathematical learning can be built in the classroom according to an epistemological access stance.

Visualisation allows an approach to mathematics learning, where the specifics of the South African school landscape are taken care of, but also, as I will argue below, it touches on some general issues which are of interest to all the mathematics educators in/of the community.

First, as discussed in one of the chapters in this book, in most South African (and generally in most African) classrooms mathematics is taught in the 'academic' language of learning and teaching, which is neither the teachers' nor the learners' first language and one in which learners lack proficiency. So, as described in the wonderful book by J. Adler (2011), *code switching* or its related strategies, like code mixing, code meshing and translanguaging, are the most immediate tools in the classroom for promoting meaningful teaching. Along this stream of thought, the book provides teachers with an organised set of concrete suggestions to widen and substantially empower this first mediation tool, showing how the concurrent use of a variety of semiotic resources, like gestures, diagrams and concrete manipulatives, suitably integrated with language(s), enables effective meaningful mathematical learning to develop in the classroom. In some cases, the proposed multimodal approach can be reinforced by the use of technological apps that are freely available should pupils have access to some form of technology, e.g. i-phones.[1] These enable visual features of classroom materials to be more dynamic, and consequently improve learning. Also, the aim of merging teaching and learning with the real world is considered as a tool for generating epistemological access, and is illustrated in the book by specific suggestions, for example, introducing outdoor activities useful for developing conceptual understanding of mathematics concepts.

In this way, the visualisation proposed in the book faces and hopefully overcomes the problem pointed out by Secada some years ago (Secada 1992, p. 627) that students in developing countries with limited access to 'academic language proficiency' achieve less conceptual understanding in mathematics than their more language-proficient peers.

From a more general standpoint, it is worthwhile to highlight how the methodological approach to visualisation proposed in the book suggests reflections and proposes examples that are important for mathematics education all over the world. I will discuss here two main themes concerning visualisation and epistemological access, which I think relevant in general:

i the importance of visual mathematics and of multimodality in general for brain and learning (Boaler et al., 2016);
ii the framework of teaching for a robust understanding (TRU), as proposed by A. Schoenfeld (2013).

These are the main topics where the specifics of programme epistemological access through visualisation described in the book intertwine with general streams in universal mathematical education research and could also contribute to it in a in a two-way productive interaction. Let me very briefly highlight them.

For the first theme, the paper by Boaler et al. (2016), written by experts in mathematics education and neurosciences, argues that the body is an intrinsic part of cognition and that "the parts of our brain that control perception and movement of our bodies, are also involved in knowledge representation" (Boaler et al., 2016, p. 328) and that a "greater emphasis on visual and physical mathematics will help students understand mathematics". They also claim that "despite the prevalence of the idea that drawing, visualising or working with models is low level or for young children, some of the most interesting and high-level mathematics is predominantly visual", and, moreover, that "another reason that visual mathematics should be used in schools to a greater extent is the nature of the knowledge needed for today's high-tech world".

The proposition in this book clearly goes in the direction suggested by Boaler and colleagues; the direction in which schools and teachers are urged to go is exactly the same as that towards which the visualisation programme is pushing: "The problem of mathematics in schools is it has been presented, for decades, as a subject of numbers and symbols, ignoring the potential of visual mathematics for transforming students' mathematical experiences and developing important brain pathways. The new brain research showing the importance of visual thinking should also prompt changes in the ways we view students in schools" (Boaler, 2016, p. 328).

As for the second theme, A. Schoenfeld (2016) developed a project at the Berkeley Graduate School of Education in order to answer the following question: "What are the attributes of equitable and robust learning environments – environments in which all students are supported in becoming knowledgeable, flexible, and resourceful disciplinary thinkers?" He argued that the quality of a learning environment depends on the extent to which it provides opportunities for students, along the following five dimensions:

***Dimension 1**: The richness of disciplinary concepts and practices available for learning (Content).*

The extent to which classroom activity structures provide opportunities for students to become knowledgeable, flexible and resourceful disciplinary thinkers. Discussions are focused and coherent, providing opportunities to learn disciplinary ideas, techniques and perspectives, make connections and develop productive disciplinary habits of mind.

***Dimension 2**: Student sense-making and "productive struggle" (Cognitive Demand).*

The extent to which students have opportunities to grapple with and make sense of important disciplinary ideas and their use. Students learn best when they are challenged in ways that provide room and support for growth, with task difficulty ranging from moderate to demanding. The level of challenge should be conducive to what has been called "productive struggle".

***Dimension 3**: Meaningful and equitable access to concepts and practices for **all** students.*

The extent to which classroom activity structures invite and support the active engagement of all of the students in the classroom with the core disciplinary content being addressed by the class. Classrooms in which a small number of students get most of the "air time" are not equitable, no matter how rich the content: all students need to be involved in meaningful ways.

***Dimension 4**: Means for constructing positive disciplinary identities through presenting, discussing and refining ideas (Agency, Ownership and Identity).*

The extent to which students are provided opportunities to "walk the walk and talk the talk" – to contribute to conversations about disciplinary ideas, to build on others' ideas and have others build on theirs – in ways that contribute to their development of agency (the willingness to engage), their ownership over the content and the development of positive identities as thinkers and learners.

***Dimension 5**: The responsiveness of the environment to student thinking (Formative Assessment).*

The extent to which classroom activities elicit student thinking and subsequent interactions respond to those ideas, building on productive beginnings and addressing emerging misunderstandings. Powerful instruction "meets students where they are" and gives them opportunities to deepen their understandings.

This programme of visualisation and epistemological access is coherent with Schoenfeld's ideas of teaching for robust understanding, and it would be interesting for the researchers and teachers engaged in the projects summarised in this book to analyse and report how and at what extent their work with visualisation corresponds with the issues of the TRU framework: it would be a nice contribution towards a progression in the research and for linking it to such an important framework as Schoenfield's one.

In conclusion, this book is relevant not only for the good examples it puts forward for countries like South Africa and Namibia, where visualisation as epistemological access is culturally rooted and necessary, but also as a scientific proposal that has great value within the research in mathematics education. It seems to me that, in a sense, it answers the call made by W.G. Secada some years ago to include the issue of diversity within the mainstream of mathematics education research:

> With few exceptions, work in this area was not found in mathematics education research journals, nor was it the product of mathematics educators. Such a state of affairs is both unconscionable and untenable. If the intellectual agendas of mainstream mathematics education do not (or cannot) include issues of student diversity, then the fundamental utility of mainstream efforts must be seen as suspect. Moreover, just as the differential distribution of mathematics among diverse learners can be seen as an equity issue, so, too, can the marginal status of work that deals with diversity.
>
> (Secada, 1992, p. 654)

This book is indeed a work that through concretely developing the issue of epistemological access through visualisation deals with diversity as an equity issue, and certainly not in a marginal way.

References

Adler, J. (2001). *Teaching Mathematics in Multilingual Classrooms.* Dortrecht: Kluwer Academic Publisher.

Boaler, J., Chen, L., Williams, C., Cordero, M. (2016). Seeing as understanding: The importance of visual mathematics for our brain and learning. *Journal of Applied Computational Mathematics*, 5(5), 325–331. doi: 10.4172/2168-9679.1000325.

McCrocklin, S. (2021). Mobile Penetration in South Africa. Retrieved on 06-02-2022 from https://www.geopoll.com/blog/mobile-penetration-south-africa/.

Morrow, W. (2009). *Bounds of democracy: Epistemological access in higher education.* Cape Town: HSRC Press.

Muller, J. (2014). Every picture tells a story: Epistemological access and knowledge. *Education As Change*, 18(2), 255–269.

Schoenfeld, A. H., & The Teaching for Robust Understanding Project. (2013). *An Introduction to the Teaching for Robust Understanding (TRU) Framework.* Berkeley, CA: Graduate School of Education. Retrieved on 06-02-2022 from http://truframework.org or http://map.mathshell.org/trumath.php.

Secada, W. G. (1992). Race, ethnicity, social class, language and achievement in mathematics. In: D. A. Grouws (Ed.), *Handbook of Research on Mathematics Teaching and Learning.* New York: MacMillan, 623–660.

1 Introduction to the Book

Marc Schäfer

Visualisation in mathematics is not only about the capacity for creating and processing physical, imaginative and virtual images of mathematical ideas and concepts; it is also about mediating mathematical meaning and understanding. It is the latter that is at the core of this book. The notion of visualisation is interrogated from a variety of perspectives and assumptions, all within the context of the classroom. These perspectives range from learning processes to teaching through a variety of media such as teaching and learning aids, bodily actions, language and technology. As much as visualising is an internal process, this book argues that this process happens in a specific context and should thus be mediated accordingly. It is interesting that the English word 'see' or 'seeing' have different meanings in this regard. The meaning of 'seeing' that naturally comes to mind refers to the obvious process of perceiving and identifying objects and one's surroundings through our sense of sight, i.e. with our eyes, via a complex neurological process. The moment light meets the retina via the cornea, pupil and lens we begin the process of seeing.

There are, however, many other meanings of the word 'see'. For example, to 'see' can refer to the moment when one suddenly understands a particular idea or concept. It is the moment when 'the penny drops'. A typical exclamation would be, 'Yes, I see what you mean'...without necessarily referring to what one sees through one's eyes, but rather when there is clear understanding in our minds of that particular moment. Another meaning of 'see' could refer to receiving a student who has made an appointment. A typical explanation would be, 'Yes, I can see you now'. Although seeing with our eyes is key to visualisation, this book takes a broader view of visualisation to incorporate other processes inherent in visualising, such as language and gestures.

As mediators of mathematical ideas, it is thus incumbent on teachers to critically interrogate visualisation processes and not make the assumption that we all see the same thing. By implication then, we see what the teacher sees or has in mind. Unfortunately, we do not have access to physically observe what goes on in our minds to discern exactly what we and others are seeing or imagining. We often rely on utterances and bodily actions from the communicator to discern

DOI: 10.4324/9781003172420-1

what they are imagining, i.e. what the communicator has created mentally. These bodily actions can form clues about what the communicator is thinking and trying to make explicit. They enable others to access what the communicator is thinking and imagining. They can also form important clues for a teacher to assess the meaning-making processes of learners.

This book argues that more attention should be given to how learners visualise and how teachers can embrace visualisation processes in their own pedagogy. In our observations, teachers use visualisation processes spontaneously and often inconsistently without giving them the attention they perhaps deserve. Using visualisation processes more strategically and consciously has the potential to not only enrich the teaching discourse, but also enable learners to access knowledge in a way that is interesting and meaningful.

This book looks at visualisation processes in mathematics from four vantage points. These form the main themes in which the individual chapters are located. They are:

- Visualisation and Pedagogy;
- Visualisation and Learning;
- Visualisation and Technology;
- Visualisation, Semiotics and Language.

The structure of the book is thus:

Chapter 1	Introduction to the Book	Marc Schäfer
Chapter 2	Towards a Theory of Visualisation	Vimolan Mudaly and Marc Schäfer
Part 1	**Visualisation and Pedagogy**	
Chapter 3	A Case for Number Sense	Ronald Griqua
Chapter 4	A Case for Fractions	Charity Ausiku
Part 2	**Visualisation and Learning**	
Chapter 5	Figural Pattern Generalisation	Duncan Samson
Chapter 6	Reasoning in Solving Word Problems	Beata Dongwi
Chapter 7	Algebraic Expressions	Sindisiwe Herbert
Part 3	**Visualisation and Technology**	
Chapter 8	Learning with *GeoGebra*	Deepak Mavani
Chapter 9	Teaching and Learning with Mobile Technologies	Clemence Chikiwa and Matthias Ludwig
Part 4	**Visualisation, Semiotics and Language**	
Chapter 10	A Case for Gestures	Cristina Sabena and Marc Schäfer
Chapter 11	A Case for Language	Clemence Chikiwa
Chapter 12	Final Word – A Synthesis	Marc Schäfer

The authors, each associated with classrooms in Southern Africa and Europe in one way or another, all have experiences in empirical research in mathematics classrooms – some more than others. The chapters are written in a style and genre that is less formal than journal articles, as each chapter relates to a uniquely personal story. Without wishing to compromise on rigour and validity of the research, each chapter is structured as a narrative that provides a brief vignette into the broader research journey of the authors as they interrogate issues around visualisation in mathematics education and epistemological access. This book agrees, and indeed argues, that epistemological access can only be assured through meaningful access to knowledge, which, according to Muller (2014, p. 257), is about "learning how to become a participant in a practice".

In the context of this book, 'practice' refers to teaching practice and the practice of learning. Each chapter provides an argument that the process of making mathematics visible through appropriate mediation goes a long way towards making mathematics accessible. This is of particular relevance in this region where access to rich teaching and learning resources largely remains elusive due to financial, logistical and socio-political constraints. The over-reliance on textbooks has further added to a reluctance to use other, often more appropriate materials and resources. This book interrogates how epistemological access, in a context of continued inequality, can be addressed through reimagining a pedagogy and classroom environment where mathematics is made more visible than it is.

This book is not meant to be didactical in the sense that it does not prescribe or even subscribe to a particular doctrine or theoretical dogma. Each chapter provides food for thought and recommends (both explicitly and implicitly) teaching and learning strategies which teachers and learners can use to make mathematics more visible and accessible than it currently is in many classrooms in the region. The chapters do not follow a set structure. The authors were encouraged to write their stories in a format that suited them and their narratives.

Chapter 2 attempts to contextualise the visualisation agenda that permeates this book, by providing a rationale for it and critically interrogating some of the definitions of visualisation and visualisation processes encountered in the research literature. It argues for as broad a definition of visualisation as possible, to include all aspects of visualising, such as recognising embodied processes and working with physical products. Inherent in the chapter is an attempt to contribute towards articulating a coherent theory of visualisation that will assist in forming frames of reference to conduct research in this field of mathematics education.

Chapters 3 and 4 are clustered in the **visualisation and pedagogy** theme with its main research focus on teachers.

Chapter 3 tells a research story of how number sense was effectively taught and mediated by a selected group of Grade 6 teachers using visual materials, in the context of a teacher intervention programme in a poverty-stricken region of South Africa.

Chapter 4 takes a critical stance on how fractions are usually taught and argues for a revised pedagogy where fractions are mediated through visual processes and objects. This research story is situated in a campus-based intervention programme for senior primary school teachers in Namibia. The main focus of this intervention programme was to interrogate how incorporating a visual approach to teaching fractions enhanced the teaching and learning of fractions, if at all.

Chapters 5, 6 and 7 are clustered in the **visualisation and learning** theme with its main research focus on learners.

Chapter 5 explores and analyses how selected high-ability Grade 9 learners made sense of number patterns that were presented to them in a pictorial context. They were individually presented with two non-consecutive terms of a linear pictorial sequence and were asked to provide as many different expressions of the *n*th term as they could. These had to be accompanied by verbal explanations and justifications.

Chapter 6 is an account of how selected groups of Grade 11 learners were observed and interacted with, as they used their own visualisation processes to solve a set of given word problems. The specific focus of this research project was on how visualisation and reasoning processes intertwined with each other in a problem-solving context.

Chapter 7 tells a research story of an after-school club that involved groups of Grade 8 learners who took selected linear algebraic expressions and invented visual patterns to represent them. The aim of this study was to bring a highly visual focus to the topic of algebraic expressions, in an attempt to make these expressions visible, and thus deepen the participants' understanding of the underlying concepts.

Chapters 8, 9 and 10 are clustered in the **visualisation and technology** theme with its main research focus on using various technologies to make mathematics more visible.

Chapter 8 interrogates how *GeoGebra* was used as a visualisation tool to learn some fundamental Euclidean geometry theorems in the context of an intervention programme. In this intervention programme the teachers developed their own applets specific to their teaching activities and implemented them in their classes. Using screen-capture software the author was able to trace and observe the learners' computer actions as they navigated the software in exploring and coming to terms with the specific Euclidean theorems.

Chapter 9 is situated in a cross-continental mobile technology project that used mobile phones in the context of set outdoor trails. It foregrounds objects in the outdoor environment and the mobile phone as two important visualisation mediators in solving authentic and interesting mathematical tasks. It tells three research stories that took place in South Africa, Namibia and Germany.

Chapters 10 and 11 are clustered in the **visualisation, semiotics and language** theme with its main research focus on gestures and language.

Chapter 10 argues that bodily actions such as gestures form an integral component of the semiotic bundle. By using case studies in Namibia and Italy, it illustrates how gestures in the teaching and learning process can enhance mathematics communication and learning. This chapter asserts that these bodily actions are manifestations of the entire body's role in generating and understanding mathematical knowledge.

Chapter 11 explores the role that the interplay between gestures, diagrams and verbal language in a selected South African multilingual class plays in providing access to knowledge when teaching mathematics. This research story involved a Grade 11 mathematics teacher teaching trigonometry.

Chapter 12 provides a brief synthesis of the book.

References

Muller, J. (2014). Every picture tells a story: Epistemological access and knowledge. *Education as Change, 18,* 255–269. doi:10.1080/16823206.2014.932256.

2 Towards a Theory of Visualisation

Vimolan Mudaly and Marc Schäfer

Introduction

One of the main objectives of this book is to deliberate and examine how the use of visualisation processes in the mathematics classroom can enhance both teaching and learning. By drawing on classroom research conducted in the Southern African region, the intention of this book is to contribute to the understanding of quality mathematics teaching in order to enhance epistemological access to mathematics at a school level. It is hoped that the context of gross inequality, under-resourced schools, unequal distribution of quality teaching and minimal support can constructively be addressed by providing research-based solutions and recommendations. Whilst its underlying purpose is to interrogate this access, this chapter also serves to briefly interrogate and search for a broad, yet deep understanding of the theoretical concepts associated with visualisation in mathematics education.

This chapter succinctly traces and outlines different definitions of visualisation in mathematics, in an attempt to analyse how these have informed past research and guided our current understanding of what it means to make mathematics visual. It also provides the rationale for the book, by emphasising a shift in the definition of visualisation from a purely product (image) perspective to a more process-oriented understanding of visualisation. In its attempt to be more all-encompassing, the chapter makes a case for a more embodied understanding of visualisation to include aspects such as body actions, semiotics, gestures and language – all integral components of a process-oriented definition of visualisation. This chapter will also briefly deliberate on the epistemological access implications of adopting a visualisation approach to teaching and learning in the context of Southern Africa, and how this approach could contribute to bridging the inequality divide that still exists in mathematics education in this region.

It can be argued that there exists a paradox in how we currently experience mathematics. For many, mathematics is unchanging, rather technical and trapped in similar routines to when human beings first began to use mathematical principles in their attempts to understand the world. For others, however, the world is changing so fast, that in education, we often struggle to keep

DOI: 10.4324/9781003172420-2

up with the transforming pedagogies and technologies that are emerging. In parallel, as Fenyvesi et al. (2014, p. 7) state, "mathematics, like all the other sciences is [also] developing so fast that it is impossible to integrate into the curriculum all the developments taking place in the field". These changes thus remind us that in these changing times of technological innovation and development, mathematics teaching and learning should be student centred and based on active learning methodologies. One strategy of enabling a student-centred teaching and learning approach is to recognise the importance of the visual and incorporate visualisation into the mathematics classroom. This is corroborated by Hutmacher's (2019, para. 2) research when he suggested that "vision is our most important and most complex sensory modality and this is mirrored in the number of studies" conducted over time. In fact, Hutmacher (2019) cites Pike et al. (2012) who state:

> [That the reason there has been extensive research on vision] is because when we interact with the world we rely more on vision than on our other senses. As a result, far more of the primate brain is engaged in processing visual information than in processing information from any of the other senses.

Arcavi (2003, p. 215) also viewed vision as being "central to our biological and socio-cultural being". Vision is researched widely in education because visual thinking is used as a lever by teachers and students to accomplish higher levels of engagement with concepts. The use of visuals allows for deeper, insightful understanding of concepts. Duval (1999, p. 3) states that "representation and visualization are at the core of understanding in mathematics". Although this is not a new idea, it lies at the core of this chapter. Bruner (1966) also articulated the importance of visuals in his theory about learning a concept. He divided the learning process into three stages, namely, the enactive, the visual and the symbolic. The enactive stage emphasises the active aspects of learning, which include being involved in and doing mathematics; for example, the learner engages by using concrete objects such as physical manipulatives to do mathematics. Manipulatives, by their very nature, are inherently visual, whether they are physical or digital (as in virtual manipulatives). The second stage involves the crafting of visual images in one's mind, which ultimately leads to the third stage – the symbolic stage – where conceptual understanding, attained through engaging in the previous two stages and using concrete objects, are assigned iconic representations. See Figure 2.1 for a visual representation of Bruner's model. This representation illustrates a continuum of learning where initially a visual stimulus is received through seeing (i.e. through sight), which is then internalised, and through a process of developing and growing insight, learning takes place.

Due to the tactile nature of manipulatives the link between using them and visualisation seems self-evident, that is, the process from sight to insight is therefore intact. This, however, is only so if the manipulatives are used

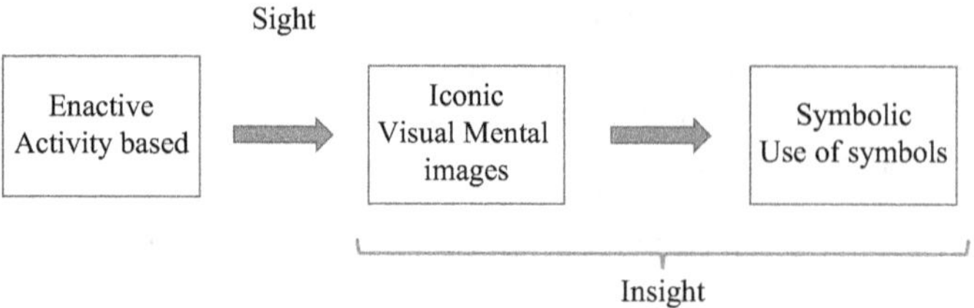

Figure 2.1 Adaptation of Bruner's (1966) model.

appropriately, usually through good mediation and support. In our current environment of technological innovation and access to a vast array of computer-generated interactive technologies, both physical and virtual manipulatives make active engagement and visualisation of concepts much more viable. The use of dynamic geometry software, for example, allows for learners to engage with a large number of mathematical concepts in a dynamic way, within a very short period of time. This would not be possible if pencil and paper methods were used. This visual approach thus enables learners to develop conceptual understanding through an interplay between sight and insight. Theme 3 in this book reports on how dynamic software can be used to enable conceptual understanding in geometry.

Visualisation and Reasoning

The National Research Council (2006, p. 3) of America asserts that "spatial thinking is based on a constructive amalgam of three elements: concepts of space, tools of representation, and processes of reasoning". This assertion aligns well with our earlier argument that visualisation – of which spatial thinking is a subset – is an intertwined outcome of product and process: it lies on a continuum from sight to insight. Reasoning is an important element in this continuum. As reasoning is all about sense making (Brodie, 2010) and drawing inferences, we argue that this is a key ingredient to visualisation processes and thus spatial thinking. When specifically referring to geometry, Jones and Tzekaki (2016, p. 114) emphasise that visualisation and spatial reasoning overlap in the sense that visualisation is about seeing, representing, transforming, generating, communicating, documenting and reflecting on visual information, whereas spatial reasoning is about making mathematical sense of these processes.

By its very nature, spatial thinking involves the manipulation (physically or mentally) of objects – which may represent mathematical concepts – through which reasoning occurs. The idea of physically or mentally manipulating objects, together with visualising and reasoning, allows the learner to "perceive, remember, and analyse the static and, via transformations, the dynamic properties of objects and the relationships between objects" (The National Research Council, 2006, p. ix). Through the innovative use of digital technologies and devices, learners are able to manipulate virtually, thereby engaging in learning visually and possibly developing their potential for perception, analysis and understanding of concepts.

Whiteley et al. (2015, p. 5) cite Tahta's (1989) seminal work on how we visualise in our minds. He proposes the idea of three 'powers' when working with space in our minds. These are:

- imagining, which involves seeing what is said;
- construing, which involves seeing what is drawn or saying what is seen; and
- figuring, which involves drawing what is seen.

The awareness of these 'powers' provides a useful framework for understanding the visual mind, because visualising involves deep mental articulation of ideas and concepts.

The underlying theme of this chapter is that visualisation occupies a subliminal space which encompasses the concepts of spatial reasoning and spatial thinking.

History of Visualisation and Research

Much has already been written about the history of visualisation research in a chapter by Presmeg (2006, pp. 205–235). We do not wish to replicate similar arguments and narratives but observe that research on visualisation is relatively new. Locatelli et al. (2010, p. 75) concur when they acknowledge that "historically, educational research has emphasised verbal learning while interest in visual learning has lagged behind".

Although formal visualisation research in mathematics might have been scarce prior to the 1980s, we can assume that visual thinking has been around ever since humans conceived of mathematical concepts, even in their most fundamental forms. The very basic activity of counting livestock must have evoked some sense of number and quantity, which would have led to the establishment of some symbolic notation of these numbers. These symbols were the foundation of formalising visuals. In his history of mathematics, Polthier (2002, p. 2) referred to a rich set of visualisations found on the copper plate engravings of Hermann Amandus Schwarz in 1865, which contained the theoretical solution to the Gergonne problem, together with a large number of other visual images. These images were there for a purpose: they were meant to *do* something for the reader. Perhaps they supplemented or complemented the theoretical proof. Perhaps the theoretical proof could not stand on its own and the images were meant to increase insight and improve understanding of the original problem.

Zimmermann and Cunningham (1991, p. 2) also recognised visualisation as an old invention or practice. They cited Ivan Rival (1987) in his book *Picture Puzzling*, who wrote that:

> Diagrams are, of course, as old as mathematics itself. Geometry has always relied heavily on pictures, and, for a time, other branches of mathematics did too.

> Even Isaac Newton… did not actually prove [the] fundamental theorems…. Had you asked him to justify them, he would likely have presented an argument that, though compelling, was loose and depended heavily on pictures.

Kadunz and Yerushalmy (2015, p. 463) specifically list the research conducted in visualisation related to education. They refer to the work of Norma Presmeg (1986, 1994, 1997), Theodore Eisenberg (1994) and Abraham Arcavi (2003), who all built on the work by David Tall (1991) who "constructed computer-based programs which made visualizations of concepts such as gradients, integrals and solutions to differential equations" because he felt that "students generally have very weak visualisation skills in mathematics and the use of computer-based programs could be one way to improve the situation" (Bråting & Pejlare, 2008, p. 1).

Of course, the use of visualisation in mathematics has also attracted criticisms, as noted by Bråting and Pejlare (2008, p. 2). In the early 19th century, for example, criticisms were levelled against visual representations with typically poor and weak constructions. Mancosu (2005, pp. 13–26) explores the current resurgence in interest in visualisation against the secondary role that it played in the past. Mancosu (2005, p. 13) found that:

> … the development of mathematics in the nineteenth century had shown that mathematical claims that seemed obvious on account of an intuitive and immediate visualisation turned out to be incorrect on closer inspection.

He further states that:

> in short, visualization seemed to lose its force in the context of justification while being allowed in the context of discovery and as something that simplifies cognition (but cannot ground it).
>
> (Mancosu, 2005, p. 14)

We emphasise that in a teaching and learning context, where visualisations are often not immediately obvious and do not stand on their own, they should be mediated appropriately and effectively. Diagrams and visuals play different roles and we argue that it is important that the consumer of these visuals understands the roles of these visuals to make sense of why they are there and what their significance is.

Since the early 1990s there has been a keen interest in research in the area of visualisation in mathematics. Zimmermann and Cunningham (1991, p. 1) claim that "to a significant degree, this renaissance [was] being driven by technological developments". As will be demonstrated elsewhere in this book, technology enables mathematics to become highly visual. Users of this technology are able to dynamically manipulate mathematical concepts and representations to explore and engage with the complexities of these concepts in ways that static pen-and-paper visuals cannot. Our visualisation model later in this chapter explains how visual images are converted into insightful images for deeper understanding.

At this stage it suffices to say that "visualisation appears to be a vivid part of research within mathematics education" (Kadunz & Yerushalmy, 2015, p. 466).

Challenges When Using Visualisation

Like most aspects of learning and knowledge acquisition, visualisation also has its challenges. It is easy to be misled by a diagram that 'looks' like something that it does not really represent. Often mistakes are made by assuming properties that are not given in the question. A quadrilateral that looks like a square is not one unless we are explicitly told that the sides are equal and at least one internal angle is equal to a right angle. Initial poor assumptions will result in poor solutions.

Historically, visualisation did not initially receive favourable reviews. The Stanford Encyclopedia of Philosophy (2020) cites Russell (1901) who pronounced that "in the best books there are no figures at all". But by the late 20th century this had changed, as can be seen by the quotation from Mancosu (2005) earlier in this chapter.

Also, visualisation is highly dependent on a student's ability to imagine, conceptualise, problem solve and recognise patterns (Haas, 2003). Imagination comes with practice and is reliant on the student's understanding of their prior knowledge. Mental manipulations can only be effective if the student is very familiar with previous concepts. Being visual also requires the ability to look at concepts holistically rather than simply remembering isolated facts. Solving problems using visualisation requires creative thinking and an expansive array of solution techniques. Finally, visual thinkers recognise patterns mentally quite easily. This ability comes with experience – to 'see' similarities and differences from a mathematical perspective.

The final challenge that should be mentioned is that visuals usually contain information that needs to be interpreted by the student. This data can only be useful and manipulated mentally if the students have practised this skill.

Towards a Definition of Visualisation

In attempting to arrive at a suitable working definition of visualisation, we present a few definitions. We do not claim that any one of these is privileged over or more important than the others. We recognise that as the research in visualisation evolved, the definitions of visualisation also evolved and were adapted to suite the prevailing narratives and research findings. We suggest that as the visualisation discourse has become more nuanced over time – from a discourse that initially only recognised visuals in terms of product such as objects, images and diagrams, to a discourse that now incorporates visual and mental processes – an all-encompassing definition should embrace the entire spectrum of visualisation.

Yilmaz and Argun (2018, p. 41) argue that visualisation is a tool for conceptual understanding. They suggest that "visualisation is a strong tool for searching mathematical problems, giving meaning to mathematical concepts and the relationships between them".

Özkan et al. (2018, p. 354) feel that visualisation is a means to make an object, body, substance, product or behaviour, action, process or an activity visible. They see it as a process of making the unseen, seen. One only needs to imagine how a learner, for example, visualises the idea of adding two and three. There are so many ways that this can be 'seen' in the mind. Mancosu (2005, p. 13) assumed a broader conception of visualisation to "include both visualisation by means of mental images as well as visualisations by means of computer-generated images or images drawn on paper, e.g. diagrams etc.".

In examining the different definitions and conceptions of visualisation, it goes without saying that the common thread that goes through all of them involves seeing a mental or physical image, visual or object in some way or another. Hershkowitz (1989, p. 75), for example, classified visualisation as "the ability to represent, transform, generate, communicate, document, and reflect on visual information". This definition still considered the idea of acting on visually available information to "represent, transform, generate, communicate, document, and reflect". Zimmermann and Cunningham (1991, p. 1) took the term visualisation to "describe the process of producing or using geometrical or graphical representations of mathematical concepts, principles or problems, whether hand drawn or computer generated". This definition was also rather restrictive and considered visualisation only as a process of generating a physical representation of a mathematical concept. This idea, however, evolved and started to include internal, mental processes as integral to visualisation processes in the quest for deeper understanding of mathematical concepts.

Zazkis et al. (1996, p. 441) provided a comprehensive definition of visualisation that captured the idea of internal and external processes quite comprehensively. They argued that external processes rely heavily on the senses (mainly sight) and internal processes depend on perception and insight.

Visualisation is an act in which an individual establishes a strong connection between an internal construct and something to which access is gained through the senses. Such a connection can be made in either of two directions. An act of visualisation may consist of any mental construction of objects or processes that an individual associates with objects or events perceived by her or him as external. Alternatively, an act of visualisation may consist of the construction – on some external medium such as paper, chalkboard or computer screen – of objects or events that the individual identifies with object(s) or process(es) in her or his mind.

The definition begins to allude to the importance of mental constructions of objects and processes, showing the movement of sight (external objects) to insight (internal processes). This evolution in the definition culminated in the widely used and cited definition by Arcavi (2003, p. 217) when he stated that:

> Visualisation is the ability, the process and the product of creation, interpretation, use of and reflection upon pictures, images, diagrams, in our minds, on paper or with technological tools, with the purpose of depicting and communicating information, thinking about and developing previously unknown ideas and advancing understandings.

In support of his own definition, Arcavi (2003) further suggested that "visualisation is no longer related to the illustrative purposes only, but is also being recognised as a key component of reasoning (deeply engaging with the conceptual and not merely the perceptual), problem solving, and even proving" (p. 235).

This aligns well with the purpose of this book which presents numerous case studies of how product- and process-orientated visualisation practices were used and harnessed in learning environments that attempted to enhance conceptual understanding of mathematics. All the studies reported on in this book commit to the notion that visualisation is more than just about seeing and perceiving – it is an intricate symbiotic process that includes the intertwinement of sight and insight.

Theoretical Frameworks that Support Visualisation

In general, we create knowledge, explain, predict and understand new ideas by using appropriate theories. Often, these theories allow us to extend our understanding by challenging existing perspectives of a particular phenomenon. Theoretical frameworks become the backbone of a study and support the research from the conception stage to final completion. Collins and Stockton (2018, p. 2) state that "a theory is a big idea that organises many other ideas with a high degree of explanatory power". They further claim that "a theoretical framework is the use of a theory (or theories) in a study that simultaneously conveys the deepest values of the researcher(s) and provides a clearly articulated signpost or lens for how the study will process new knowledge" (Collins & Stockton, 2018, p. 2). There are, however, a number of theories that have been used in mathematics research and these can easily be found using literature searches.

In the absence of a single theory of visualisation, we wish to briefly consider a few possible theoretical frameworks that could prove useful in research involving visualisation, and could possibly contribute towards a more rigorous theoretical deliberation on the underpinning of visualisation research in mathematics education. Whilst the four theories presented here (albeit rather briefly) may not appear to be interconnected, they are theories that we choose to promote in order to encourage researchers in the field to explore different lenses when attempting to understand visual phenomena in the context of mathematics education.

Özkan et al.'s (2018, p. 354) statement that "visualisation is an effective method for students to internalise concepts" is a useful platform to consider possible frameworks that may underpin visualisation research and deliberations, remembering of course that visualisation processes involve both internalising and externalising actions. The following theoretical frameworks are considered:

- the iterative visualisation thinking cycle
- commognition
- embodiment
- enactivism

Iterative Visualisation Thinking Cycle

In his work, Mudaly (2021) proposes a four-stage visualisation thinking cycle:

- Stage 1: doing stage
- Stage 2: seeing stage
- Stage 3: symbolising stage
- Stage 4: applying stage

Figure 2.2 illustrates these cycles.

Stage 1

Mudaly (2021, p. 57) argues that in Stage 1 a learner reacts to a visual mediator or stimulus such as a manipulative or physical object and starts to engage

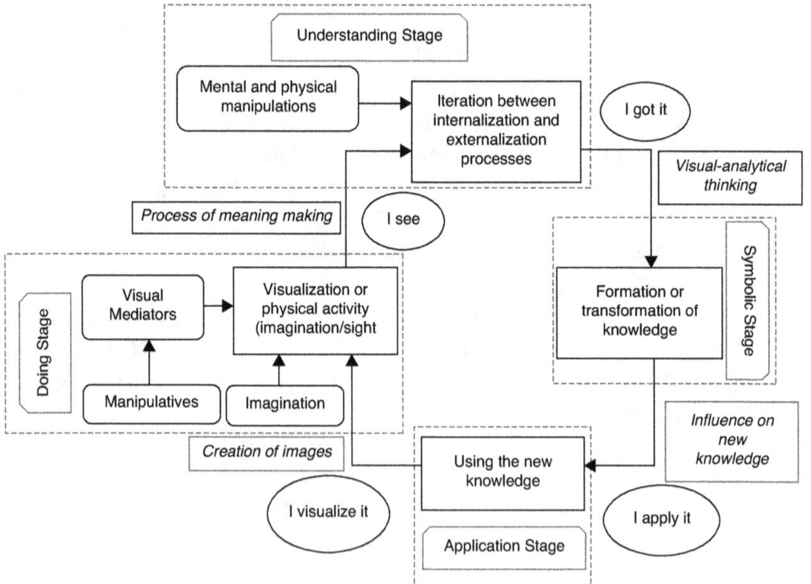

Figure 2.2 Iterative visualisation thinking cycle (Mudaly, 2021, p. 57).

with this object. Thus begins a visualisation activity with some kind of active engagement. This could include any physical activity related to a particular concept that is under consideration. These activities could include, for example, drawing, reading, manipulating, listening or engaging with some technology. They could also involve mental activities such as imagining, recalling or reflecting. In this stage the learner is able to 'see' the concept unfolding ('I visualise it').

Stage 2

In this 'understanding stage' the learner is actively engaged in trying to make sense of a concept, often by trial and error, and using iterations of external and internal processes. These iterations of 'acting' on a stimulus or mediator ('doing') and then reflecting ('thinking') on it can result in an iterative process of doing and thinking. These mental and physical manipulations are often subtle and occur almost simultaneously. As the learner continues to engage, a level of understanding results, by associating the new information and experience with previously acquired knowledge. This then leads to the next stage – the seeing stage – where the learner begins to understand the concept or the manipulation. The use of dynamic software, for example, enhances this stage of visualisation and enables a certain level of understanding. With the new understanding, further analysis ensues to establish an even higher level of understanding. This thinking will continue until the requisite level of understanding is attained ('I see').

Stage 3

With deep mental understanding of the concept the learner then needs to convert it into a symbolic, mathematical form. This is the 'symbolic stage' where understanding results in the formation of new knowledge or the transformation of existing knowledge. At this stage the learner should, for example, be able to produce a proof ('I got it').

Stage 4

The final stage is the 'application stage' where the new knowledge is used to explain and solve problems in the contexts presented. Once attained, the cycle may begin again with a new concept ('I apply it').

Commognition

Central to the notion of commognition is the idea that visualisation is a form of communication – communication with oneself in making sense of a concept, and communication with others in conveying certain data and information. Sfard (2007, p. 567) presented a useful interpretive framework for the study of learning, which she termed 'commognition'. This "is grounded in the assumption that thinking is a form of communication and that learning mathematics is tantamount to modifying and extending one's discourse" (Sfard, 2007, p. 567). Sfard (2007, p. 571) further explained that "human thinking can be regarded as the individualised form of the activity of communicating, that is, as communication with oneself. This self-communication does not have to be in any way audible or visible, and it does not have to be in words". Sfard (2008, p. xvii) defined the term 'commognition' as a combination of communication and cognition, and stresses that interpersonal communication and individual thinking are two facets of the same phenomenon.

It is important to examine how children think mathematically when learning concepts or when solving problems, and Sfard's theory offers us an opportunity to delve deep into the minds of children by getting them to talk about what they think, as well as to encourage learners to illustrate the actual mental pictures they 'see'. Sfard (2008, p. 133) saw visual mediators as "visible objects that are operated upon as a part of the process of communication". In communicating with the self, these visible objects are often associated with mental images. The mention of the number two, for example, immediately brings forth an image of '2'. In a specific context it might be that a mental image of two objects may appear. For advanced thinking, even the gradient of 2 can be visualised. So, Sfard's theory of commognition has far greater import in the field of visualisation than has previously been acknowledged.

Embodiment

At the heart of embodied theories of cognition (also sometimes referred to as 'grounded' or 'situated' cognition theories) is the notion that the body, with many of its actions, and cognitive processes, such as learning, are intricately linked. As Glenberg (2008, p. 355) noted, "there is a close connection between the body and education".

According to Kiefer and Trumpp (2012), cognition was traditionally assumed to involve neuro-cognitive systems only, which were divorced from processes that involved sensory-motor activities and information. This classical view is being challenged more and more with the proposal that there are "close links between the sensory and motor brain systems on the one hand, and cognition on the other hand" (p. 16).

Kiefer and Trumpp (2012) cited an interesting example from their research which has relevance to our visualisation narrative. They asked the question of how learners know that the number 6 is greater than the number 4, bearing in mind that numbers are essentially abstract concepts. They suggested that analysing number magnitude involves accessing a mental number line. This, in essence, translates into visualising a mental number line and determining the relative magnitude of these numbers by their positions in relation to each other. A further argument in relation to visualisation is that numbers are grounded in bodily actions. As Kiefer and Trumpp (2012, p. 19) argue, number comparisons are "strongly influenced by the finger counting habits" of individuals. They thus conclude that number concepts "appear to be embodied in both visuo-spatial and action-related representations".

Furthermore, Alibali and Nathan (2012) argue that mental processes such as thinking, reasoning and sense making

> ... are mediated by bodily based systems, including body shape, movement, and scale; motor systems including the neural systems engaged in action planning; and the system involved in sensation and perception.
>
> (p. 248)

Enactivism

Enactivism is an interesting and useful theoretical underpinning which is gaining traction in mathematics education. It puts the role of the observer at the centre and asserts that "anything said is said by an observer" (Maturana & Varela, 1992, p. 8). Maturana, one of the foundational thinkers in enactivism, implied by this that "[what] is said can under no circumstances be separated from the person saying it" (Maturana & Poerksen, 2004, p. 26). Maturana further asserts that what exists independently of the observer is necessarily a matter of belief and not of knowledge, because to see something always requires someone who sees it (Maturana & Poerksen, 2004, p. 28). From an enactivist perspective, if we translate this to visualisation, it means that for someone to see something or perceive something, that person needs to see this something. We can thus not assume that we all see the same thing. The question 'Is what I see the same as what you see?' is thus a very legitimate question and has important implications for a teacher who makes use of visuals and images.

In enactivism, the mind and the body are inseparable – very much an embodied concept (as alluded to above). The bringing together of the mind and the body is, according to Varela et al. (1991), achieved through a process of open-ended reflection. Di Paolo et al. (2007, p. 12) suggest that mind and body form part of a living system, such as a learner composed of various "autonomous layers of self-coordination and self-organisation". This allows the learner to interact with the world in sense making such as learning. We argue that visualisation is a powerful means to interact with the world and mediate between body and mind in order to make sense of what is observed.

In terms of thinking about the learning process, Kieren et al. (1995, p. 2) assert that enactivism refers to a given situation where we are called to position ourselves

> ... and view cognition not best seen in terms of its products nor its mental structure, but in terms of action in, perhaps better yet, as living in the world of significance with others.

Some of the more recent researchers who used enactivism as their theoretical lens include Dongwi (2018, p. xx), who researched how learners used visualisation processes to solve geometry word problems. She used enactivism to inform both her methodological and analytical frameworks as it provided her "with the language to discuss the links between visualisation and reasoning during word problem solving". Reid (2014), in his study on the processes of visualisation in the learning of number operations by children, also employed an enactivist perspective for his theoretical and methodological framing. Froese (2015) added that enactivism provides a suitable interpretative framework for explaining the finding that emotional networks are among the most widely connected in the brain and help us to better understand the role of 'nonstandard' pathways to visual perception (p. 3).

Epistemological Access

The concept of epistemological access in Southern Africa has its roots in the early 1990s when the education philosopher Wally Morrow (1994) argued that access to knowledge and 'the discipline', in the context of transforming higher education, needed to go beyond simple access to the institution. Reflecting on Morrow's contribution to educational thinking in the region, Muller (2014, p. 257) argued that gaining access to knowledge is about "learning how to become a participant in a practice" – it is about gaining "meaningful access to the goods of the university". This was, and still is, a very pertinent concern in a region with a long history of racial discrimination and oppression, where only certain so-called race groups were guaranteed this access. The question of how this access can be achieved, even in the current climate of critical social, political and decolonialist reflections and narratives, remains as elusive as ever. Is it a question of hard work, of good teaching and research, of transformation, of curriculum and policy reform, of economics and affordability, or is it a question of entitlement? Perhaps the answer lies in all of them.

We argue that the epistemological access narrative should percolate to the schooling arena, as it is already there that meaningful access is denied to so many learners. This is not only a regional concern, but applies across the entire globe. A perusal of recent Trends in International Mathematics and Science Study (TIMSS) and Progress in International Reading Literacy Study (PIRLS) results evidences the iniquitous achievements in mathematics and reading across the globe and attests to less than equal access to meaningful schooling for many learners. It is thus clear that access to any school does not necessarily imply access to a meaningful education.

The same can be argued on a more micro level such as the mathematics classroom. If access to good mathematics education cannot be guaranteed, we are then faced with a social justice issue. Any classroom is thus a terrain where social justice issues play out on a daily basis, and need to be interrogated. Bond and Chernoff (2015, p. 27) note that "[b]ecause the classroom itself is one of the primary sources of the socialisation that shapes social inequality, an uncritical pedagogy serves only to enforce existing systems of dominance and inequity". We thus argue that with epistemological access comes access to mathematics classrooms where quality pedagogies are practised and learning environments created, which enable access to the discipline and its practices in such a way that every learner can indeed become a participant in those practices.

Conclusion

As each researcher in his/her chapter in this book deliberates with the implications of how visualisation processes can be incorporated in the mathematics classroom, the significance of the findings speak to how teaching can be improved so that epistemological access to mathematics becomes less elusive in this region. We argue in each chapter that adopting a strong visualisation agenda in the mathematics classroom enables strong and meaningful mathematical mediation, and thus provides practical and implementable ideas for improving mathematics teaching and learning in the region.

References

Alibali, M. W., & Nathan, M. J. (2012). Embodiment in mathematics teaching and learning: Evidence from learners' and teachers' gestures. *Journal of the Learning Sciences, 21*(2), 247–286.

Arcavi, A. (2003). The role of visual representations in the learning of mathematics. *Educational Studies in Mathematics, 52*(3), 215–241.

Bråting, K., & Pejlare, J. (2008). Visualizations in mathematics. *Erkenntnis, 68*(3), 345–358. https://doi.org/10.1007/s10670-008-9104-3

Bond, G., & Chernoff, E. J. (2015). Mathematics and social justice: A symbiotic pedagogy. *Journal of Urban Mathematics Education, 8*(1), 24–30.

Brodie, K. (2010). *Teaching Mathematical Reasoning in Secondary School Classrooms.* New York: Springer.

Bruner, J. S. (1966). *Toward a Theory of Instruction.* Cambridge: Harvard University Press.

Collins, C. S., & Stockton, C. M. (2018). The central role of theory in qualitative research, *International Journal of Qualitative Methods, 17*(1–10).

Di Paolo, E., Rohde, M., & De Jaegher, H. (2007). Horizons for the enactive mind: Values, social interaction and play. In J. Stewart, O. Gapenne, & E. Di Paolo (Eds.), *Enaction: Towards a New Paradigm for Cognitive Science* (pp. 33–87). Cambridge, MA: MIT Press.

Dongwi, B. L. (2018). *Examining Mathematical Reasoning Through Enacted Visualisation.* Unpublished Doctoral thesis. Rhodes University, Grahamstown.

Duval, R. (1999). Representation, vision and visualization: Cognitive functions in mathematical thinking. Basic issues for learning. In F. Hitt, & M. Santos, (Eds.), *Proceedings of the Twenty First Annual Meeting of the North American Chapter of the International Group for the Psychology of Mathematics Education. Vol. I* (pp. 3–26). Columbus, OH: ERIC Clearinghouse for Science, Mathematics, and Environmental Education.

Eisenberg, T. (1994). On understanding the reluctance to visualize. *Zentralblatt fuer Didaktik der Mathematik, 26*(4), 109–113.

Fenyvesi, K., Téglási, I. O., & Szilágyi, I. (2014). Adventures on paper – not only for the math class! In K. Fenyvesi, I. O. Téglási, & I. P. Szilágyi (Eds.), *Adventures on Paper:*

Math-Art Activities for Experience-centered Education of Mathematics. Hungary: Eszterházy Károly College.

Froese, T. (2015). Enactive neuroscience, the direct perception hypothesis, and the socially extended mind 1. *Behavioral and Brain Science, 1*, 6.

Glenberg, A. M. (2008). Embodiment for education. In *Handbook of Cognitive Science: An Embodied Approach* (p. 355–372). Amsterdam: Elsevier. https://doi.org/10.1016/B978-0-08-046616-3.00018-9

Haas, S. C. (2003). Algebra for gifted visual-spatial learners. *Gifted Education Communicator 34*(1), 30–31; 42–43. https://citeseerx.ist.psu.edu/viewdoc/download?doi=10.1.1.132.6634&rep=rep1&type=pdf

Hershkowitz, R. (1989). Psychological aspects of learning geometry. In P. Nesher, & J. Kilpatrick (Eds.), *Mathematics and Cognition: A Research Synthesis by the International Group for the Psychology of Mathematics Education* (pp. 70–95). Cambridge: Cambridge University Press.

Hutmacher F. (2019). Why is there so much more research on vision than on any other sensory modality? *Frontiers in Psychology, 10*, 2246. https://doi.org/10.3389/fpsyg.2019.02246

Jones, K., & Tzekaki, M. (2016). Research on the teaching and learning of geometry. In A. Gutiérrez, G. Leder, & P. Boero (Eds.), *The Second Handbook of Research on the Psychology of Mathematics Education: The Journey Continues* (pp. 109–149). Rotterdam: Sense Publishers. http://dx.doi.org/10.1007/978-94-6300-561-6_4

Kadunz, G., & Yerushalmy, M. (2015). Visualization in the teaching and learning of mathematics. In S. J. Cho (Ed.), *The Proceedings of the 12th International Congress on Mathematical Education* (pp. 463–467). https://doi.org/10.1007/978-3-319-12688-3_41

Kiefer, M., & Trumpp, N. M. (2012). Embodiment theory and education: The foundation of cognition in perception and action. *Trends in Neuroscience and Education, 1*, 15–20.

Kieren, T. E., Calvert, L. G., Reid, D. A., & Simmt, E. S. M. (1995). Coemergence: Four enactive portraits of mathematical activity. *Paper presented at the Annual Meeting of the American Educational Research Association* (pp. 1–35). San Francisco, CA.

Locatelli, S., Ferreira, C., & Arroio, A. (2010). Metavisualization: An important skill in the learning chemistry. *Problems of education in the 21st Century, 24*, 75–83.

Mancosu, P. (2005). Mathematical reasoning and visualization. In P. Mancosu, F. Jørgensen, & S. A. Pedersen (Eds.), *Visualization, Explanation and Reasoning Styles in Mathematics* (pp. 13–26). New York: Springer.

Maturana, H. R., & Poerksen, B. (2004). *From Being to Doing: The Origin of the Biology of Cognition.* Heidelberg: Carl-Auer Verlag.

Maturana, H., & Varela, F. (1992). *The Tree of Knowledge: The Biological Roots of Human Understanding.* Boston, MA: Shambhala Press.

Morrow. (1994). Entitlement and achievement in education. *Studies in Philosophy and Education, 13*(1), 33–47.

Mudaly, V. (2021). Visualizing as a means of understanding in the Fourth Industrial Revolution environment. In J. Naidoo (Ed.), *Teaching and Learning in the 21st Century: Embracing the Fourth Industrial Revolution* (pp. 1–10). Leiden: Brill/Sense Publishers.

Muller, J. (2014). Every picture tells a story: Epistemological access and knowledge. *Education As Change, 18*(2), 255–269.

Özkan, A., Arikan, E. E., & Özkan, E. M. (2018). A Study on the Visualization Skills of 6th Grade Students, *Universal Journal of Educational Research, 6*(2), 354–359. https://doi.org/10.13189/ujer.2018.060219

Polthier, K. (2002). Visualizing mathematics—Online. In C. P. Bruter (Ed.), *Mathematics and Art* (pp. 29–42). Berlin: Springer.

Presmeg, N. (1986). For the learning of mathematics. *JSTOR 6*(3), 42–46. https://www.jstor.org/stable/40247826?seq=1

Presmeg, N. C. (1994). The role of visually mediated processes in classroom mathematics. *Zentralblatt fuer Didaktik der Mathematik, 26*(4), 114–117.

Presmeg, N. C. (1997). Generalization using imagery in mathematics. In L. English (Ed.), *Mathematical Reasoning: Analogies, Metaphors and Images* (pp. 299–312). London: Lawrence Erlbaum Associates, Inc. https://doi.org/10.4324/9780203053485

Presmeg, N. (2006). Research on visualization in learning and teaching mathematics. In A. Gutierréz, & P. Boero (Eds.), *Handbook of Research on the Psychology of Mathematics Education: Past, Present and Future.* Rotterdam: Sense Publishers.

Reid, D. A. (2014). The coherence of enactivism and mathematics education research: A case study. *AVANT, V*(2), 137–172.

Sfard, A. (2007). When the rules of discourse change, but nobody tells you: Making sense of mathematics learning from a commognitive standpoint. *The Journal of the Learning Sciences, 16*(4), 567–615.

Sfard, A. (2008). *Thinking as Communicating: Human Development, the Growth of Discourses, and Mathematizing.* Cambridge: Cambridge University Press.

Stanford Encyclopedia of Philosophy. (2020). The epistemology of visual thinking in mathematics. https://plato.stanford.edu/entries/epistemology-visual-thinking/

Tall, D. (1991). Recent developments in the use of the computer to visualize and symbolize calculus concepts. *The Laboratory Approach to Teaching Calculus, 20*, 15–25.

The National Research Council. (2006). *Learning to Think Spatially.* Washington, DC: The National Academies Press.

Varela, F. J., Thompson, E., & Rosch, E. (1991). *The Embodied Mind: Cognitive Science and Human Experience.* Cambridge, MA: The MIT Press.

Whiteley, W., Sinclair, N., & Davis, B. (2015). What is spatial reasoning? In B. Davis and the Spatial Reasoning Study Group (Eds.), *Spatial Reasoning in the Early Years: Principles, Assertions, and Speculations.* New York: Routledge.

Yilmaz, R., & Argun, Z. (2018). Role of visualization in mathematical abstraction: The case of congruence concept. *International Journal of Education in Mathematics, Science and Technology, 6*(1). https://doi.org/10.18404/ijemst.328337

Zazkis, R., Dubinsky, E., & Dautermann, J. (1996). Coordinating visual and analytic strategies: A study of students' understanding of the group D_4. *Journal for Research in Mathematics Education, 27*(4), 435–457. https://doi.org/10.2307/749876

Zimmermann, W., & Cunningham, S. (1991). Editor's introduction: What is mathematical visualization? In W. Zimmermann, & S. Cunningham (Eds.), *Visualization in Teaching and Learning Mathematics* (pp. 1–8). Washington, DC: Mathematical Association of America.

Part 1

Visualisation and Pedagogy

3 A Case for Number Sense

Ronald Griqua

Introduction

The South African Department of Education report regarding the poor numeracy skills of its Grade 6 learners prompted me to explore alternative teaching strategies which could enhance the number sense of learners. According to the Department of Basic Education of the Republic of South Africa [DBE RSA] (2014), the Annual National Assessments (ANA) which were written in South Africa from 2012 to 2014 have flagged the poor numeracy skills of Grade 6 learners nationally. The average ANA scores of Grade 6 learners in the Northern Cape Province, from 2012 to 2014, are sadly below the acceptable target of 50%. For the John Taolo Gaetsewe (JTG) district, the average marks for Grade 6 learners for mathematics was 33.5%, which is a real cause for concern. The average marks for Grade 6 learners in the Northern Cape Province for the ANA Mathematics tests in 2012, 2013 and 2014 are shown in Table 3.1 (DBE RSA, 2014).

Although Table 3.1 indicates an improvement in the learner performance over the three consecutive years, the average provincial scores remain below the acceptable achievement level of 50% or more (DBE RSA, 2014). Table 3.2 shows that in the Northern Cape, the JTG district achieved the lowest average percentage marks for Grade 6 learners in the ANA Mathematics tests in 2013 and 2014 (DBE RSA, 2014):

The JTG district lies in a rural area with mixed socio-economic conditions, ranging from villages with minimal resources to conditions of abject poverty. These conditions impact negatively on the access that learners have to quality mathematical knowledge, due to the lack of resources in the schools of the district.

Table 3.1 The average percentage marks of Grade 6 learners in the Northern Cape in the 2012, 2013 and 2014 ANA Mathematics tests

Province	*2012*	*2013*	*2014*
Northern Cape	23.8%	35.6%	39.3%

DOI: 10.4324/9781003172420-4

Table 3.2 The average percentage marks for Grade 6 learners in the 2013 and 2014 ANA Mathematics tests for districts in the Northern Cape

Year	*John Taolo Gaetsewe (%)*	*Frances Baard (%)*	*Namakwa (%)*	*Pixley ka Seme (%)*	*ZF Mgcawu (%)*
2013	31.2	38.3	42.3	32.2	36.1
2014	35.8	43.5	42.9	35.8	37.4

Against this backdrop, a research study was undertaken to examine selected Grade 6 teachers' experiences and perceptions of the use of visualisation processes in their pedagogy when they taught number sense, after participating in an intervention programme. The research study was framed as a case study that was grounded within the interpretive paradigm. The study was located in classrooms where the participating teachers promoted active learning, after taking part in an intervention programme. A constructivist theoretical underpinning was adopted. The study employed a mixed method approach, where qualitative data was collected through observations and interviews, while quantitative data was collected through a series of pre- and post-tests.

At the heart of the study was the Visualisation Intervention Programme (VIP), which involved seven Rhodes University Mathematics Education Project (RUMEP) teachers within the JTG district of the Northern Cape Province. The content of the VIP was informed by initially working with five teachers of well-resourced schools within the Northern Cape who made interesting use of visualisation processes and manipulatives to teach number sense. With the assistance of these five teachers – the collaboration team – the VIP was then implemented by seven selected RUMEP teachers to investigate the role of visualisation processes in the teaching of number sense understandings.

The challenge for these seven participating teachers was to engage in a revised pedagogy that "best enabled epistemological access in schools through their teaching practices" (Lotz-Sisitka, 2009, p. 75). This research study was therefore conducted to investigate whether the use of visualisation processes in the teaching of number sense understandings could provide epistemological access to mathematics education and thereby improve learners' performance within the classrooms of the participating teachers.

This chapter focuses on the crafting and implementation of the VIP, the connection between visualisation and number sense, the theoretical framework that underpinned the study, the research methodology, the qualitative and quantitative analyses of the number sense research data and the findings of the research study. The chapter concludes with recommendations regarding the use of visualisation processes in the teaching of number sense to ensure epistemological access to mathematics education. Within the context of this research study, epistemological access is enabled by the extent to which the use of visualisation processes facilitated the learners' deeper understanding of number sense themes, their fluidity and flexibility with numbers, their sense of what numbers mean, their ability to perform mental mathematics and how they look at the world and make comparisons (Gersten & Chard, 1999).

The Visualisation Intervention Programme

Background

The VIP constituted an intervention programme in the Northern Cape Province aimed at enhancing the teaching of number sense to Grade 6 learners. The content of the VIP was the result of a collaboration between teachers who are experienced with and who favour a visual approach to teaching.

Presmeg (1986) defines teachers who prefer to teach visually as being of high teaching visuality. This teaching visuality refers to "the extent to which that teacher uses visual presentations when teaching mathematics" (Presmeg, 1986, p. 43). The VIP was therefore an attempt at increasing the teaching visuality of the participating teachers in this research study, in order to investigate whether and in what manner visual teaching could enhance the teaching of number sense understandings.

The implementation of the VIP facilitated classroom observations and interviews with the participating teachers, as it provided a framework to extract the views of the participating teachers regarding the usefulness of using visualisation when teaching number sense. The VIP framed the research methodology as a case study. Key to the case was the incorporation of the views and perceptions of the participants (Tellis, 1997).

We now turn to an in-depth focus on the design of the VIP in order to link our discussion of visualisation processes in mathematics education to the teaching of number sense.

The VIP as a Teaching Strategy

The collaborating team took cognisance of the need to craft a VIP that could facilitate number sense understandings through visual teaching strategies. The focus was on creating a facilitative platform for visual, mathematical instruction. Boaler et al. (2016) suggest that neuroimaging has shown that our mathematical thinking is grounded in visual processing, even if we work on a number calculation with symbolic digits. The VIP sought to take advantage of this human predisposition for visual processing in solving mathematical tasks.

Furthermore, Boaler et al. (2016, p. 2) believe that "good mathematics teachers typically use visuals, manipulatives and motion to enhance students' understanding of mathematical concepts". The VIP advocated visual teaching strategies relevant to number sense understandings, as concretised within the different problem contexts of the learner activities. These strategies included the use of whole number cards, fractional number cards, arrays of dots, the decimal fraction board, the fraction circle, the abacus, the Gattegno chart, flard cards, the fraction wall, fraction strips, the decimal number line, the conversion chart, number line stickers, diagrammatical representations, spider diagrams, base ten blocks, the multiplication board and counters, and sticks and bands.

The VIP sought to move away from presenting mathematics as merely a numeric and symbolic subject. Instead, the intention was to teach mathematics for the development of visual understandings (Boaler et al., 2016). As such, the lessons on number sense understandings were planned with accompanying visual aids.

The learner activities were designed to facilitate a problem context for number sense understandings. These activities included oral and written questions (see Figures 3.1a and 3.1b). The oral questions served as an introduction to the lesson. The objective was to check learners' prior knowledge and to engage in some mental mathematics to prepare them for the lesson topic. The lessons were developed in four stages, namely, introduction, mental mathematics, development of the lesson and consolidation of the lesson. The aim was to do justice to the constructivist perspective that instruction should allow learners opportunities to make connections between new information and their existing cognitive structures (Ausubel, 2000).

In order for the participating teachers to properly implement the programme in this case study, they were taken through a series of workshops and one-on-one deliberations regarding the implementation of visual strategies for the teaching of four lessons related to number sense. The participating teachers were made aware that the visual teacher regards visual aids as "a stimulating reference for abstract concepts" as these aids facilitate visual thinking (Alsina & Nelsen, 2006,

The following oral questions were put to the class for a lesson of "The Nature of Numbers":

- **Give the counting number before 500.**
- **If you count in quarters, what comes after $2\frac{3}{4}$?**
- **What is halve of 0,340?**
- **What is double 25%?**

Activity 1

The digits are: 0, 1, 2, 3, 4, 5, 6, 7, 8 and 9.

Put one digit in each of the two boxes so that the answer will be as big as possible. (Do not repeat the digits.)

7 ☐☐ – 312 = ?

Explain why you chose those digits.

Figure 3.1a Example of the oral questions.
Figure 3.1b Example of the written questions.

p. 125). The challenge with preparing the teachers for the implementation of the VIP was that most of them were rooted in the conventional teaching style of approaching mathematics in a predominantly verbal, procedural and algorithmic manner. These teachers were generally non-visual practitioners as far as their mathematics classroom practices were concerned (Arcavi, 2003; Presmeg, 1986). Given the significance of the VIP in relation to the research objectives, it was important that the participating teachers bought into the vision of the planned and desired visual strategies.

Selection of the Seven Participants

My work entails the development of in-service teachers who enrol in RUMEP. These teachers work in under-resourced rural schools located within the JTG district. As mentioned, this district is an area of abject poverty. The typical struggles faced by these rural schools are "a lack of parental interest in children's education, insufficient funding from the state, a lack of resources, underqualified teachers, and multi-grade teaching" (Du Plessis & Mestry, 2019, p. 1).

I invited all the in-service teachers to a series of awareness workshops to discuss the VIP which was designed for the research study. After the workshops, I invited seven RUMEP teachers to volunteer to participate in the research, based on their level of interest and active participation during the awareness workshops. I then ran a series of VIP workshops with these seven participants where we examined and explored what it means to incorporate a visual approach to teaching number sense.

As the VIP sought to enhance the teaching of number sense, a literature review of the relationship between visualisation and number sense is necessary.

Visualisation and Number Sense

Research done with pre-service primary teacher education students found that "understanding and use of mathematics can be promoted and assisted by the development of number sense" (Kaminski, 2002, p. 133). The results of this study by Kaminski (2002) suggest that the participants developed and utilised multiple relationships with numbers after they participated in a number sense programme. They furthermore made sense of the mathematics they investigated and the participants were able to provide explanations for the results they achieved (Kaminski, 2002).

This research study made use of a framework for number sense to inform the use of visualisation processes to enhance the teaching of number sense. The number sense framework in Table 3.3 formed the basis upon which I developed an analytical framework (Table 3.4) for considering visual teaching strategies for number sense.

McIntosh et al. (1992) introduced a framework of how the ***different components of number sense*** relate to each other. See Table 3.3.

Table 3.3 Overview of a framework for number sense

Major components	*Number sense understandings*	*Number sense themes*
Knowledge of and facility with numbers	Sense of orderliness of numbers	Place value Relationship between number types Ordering numbers within and among number types
	Multiple representations for numbers	Graphical/symbolic Equivalent numerical forms (including de-composition and re-composition) Comparison to benchmarks
	Sense of relative and absolute magnitude of numbers	Comparing to physical referent Comparing to mathematical referent
	System of benchmarks	Mathematical Personal
Knowledge of and facility with operations	Understanding the effect of operations	Operating on whole numbers Operating on fractions/decimals
	Understanding mathematical properties	Commutativity Associativity Distributivity Identities Inverses
	Understanding the relationship between operations	Addition/multiplication Subtraction/division Addition/subtraction Multiplication/division
Applying knowledge of and facility with numbers and operations to computational settings	Understanding the relationship between problem context and the necessary computation	Recognise data as exact or approximate Awareness that solutions may be exact or approximate
	Awareness that multiple strategies exist	Ability to create and/or invent strategies Ability to apply different strategies Ability to select an efficient strategy
	Inclination to utilise an efficient representation and/or method	Facility with various methods (mental, calculator, pencil and paper) Facility choosing efficient number(s)
	Inclination to review data and result for sensibility	Recognise reasonableness of data Recognise reasonableness of calculation.

Source: McIntosh et al., 1992

The analytical framework (Table 3.4) was used to seek evidence of visual teaching strategies for number sense and it was informed by the need to connect classroom learning related to number sense with real-life contexts, through the use of concrete and virtual manipulatives. The teachers' use of visual teaching strategies contributed to the creation of learning environments that enabled learner access to mathematical knowledge and ensured that the learners became participants in the mathematics classroom.

It is noteworthy that research suggests that learners benefit from exposure to different modes of representation (Moskal & Magone, 2000) – an integral component of using visualisation processes. These modes of representation are constituted by the manner in which information is communicated by the teacher and used by the learner. Such modes may be visual-spatial (as in concrete manipulatives and diagrams), verbal or syntactic (as in written and spoken words) or formal notational (such as mathematical symbols) (Moskal & Magone, 2000).

Furthermore, Chrysostomou et al. (2013, p. 1) found that "spatial imagery, in contrast to the object imagery and verbal cognitive styles, is related to achievement in number sense". Their research has found a strong relationship between participants' preference for using spatial imagery in solving number sense tasks and solving these tasks successfully. A distinction is made between two types of imagery subsystems. These are termed 'object' and 'spatial' imagery subsystems. Object imagery refers to the visual processing of the properties of the object which is being visualised. Spatial imagery, in contrast, refers to visual processing of the relationships between the features of the object which is being visualised.

Research also found that a learner's use of visual-spatial imagery positively correlates with the degree to which number sense tasks are understood and solved (Chrysostomou et al., 2013). Visual-spatial imagery facilitates productive visual processing of mathematical relationships. It was discovered that the successful solution of algebraic word problems is associated with the use of visual-spatial imagery (Terao et al., 2004). An emphasis is placed on the need to study the influence of cognitive styles on the learning of mathematics in order to inform our instructional practices for the development of number sense (Chrysostomou et al., 2013; van Garderen, 2006; Hegarty & Kozhevnikov, 1999).

Research confirms that when visual-spatial representations are included as a component of instruction and learners are urged to go beyond merely visualising the problem, the mathematics problem-solving skills of learners do seem to improve (van Garderen, 2006; Hegarty & Kozhevnikov, 1999). It therefore stands to reason that the teaching of number sense should accommodate strategies that could facilitate the use of visual-spatial imagery by learners. This could promote good number sense and ultimately improve their abilities to successfully solve mathematical problems. Essentially, the teachers' use of visual-spatial representations in the mathematics classroom facilitates an environment within which learners could gain epistemological access to mathematics as they are urged to fully participate in the lesson.

Table 3.4 An analytical framework for considering visual teaching strategies for number sense

Number sense understandings	*Number sense themes*	*Visual teaching strategies*
Sense of orderliness of numbers	Place value	• The abacus can be used to demonstrate place values and regrouping of a number (SAIDE, 2007, p. 36) • The use of flard cards whereby the different digits of a number are placed behind each other is helpful in teaching place values, because when the cards are exposed the values of the different digits become apparent (SAIDE, 2007, p. 37) • The calculator game called ZAP facilitates an understanding of place values. It is played as follows: One player calls out a number for the other players to enter onto their calculator displays (e.g., 4 789). The player then says, "ZAP the 8", which means that the other players must replace the 8 with the digit 0, using one operation (i.e., to change it into 4,709). The player who is the quickest to decide on how to ZAP the given digit could call out the next number (SAIDE, 2007, p. 38) • The Gattegno chart can be used to facilitate an understanding of place value (Anghileri, 2000, p. 12)
	Relationship between number types	• The number line can be populated with a starting and end point. The learner is required to insert a number or decimal value in between at a point indicated by the teacher (McIntosh et al., 1992, p. 6)
	Ordering numbers within and among number types	• The number track, calibrated and empty number lines are used for representations which focus on the ordering of numbers and the counting sequence (Anghileri, 2000, p. 10)
Multiple representations for numbers	Graphical/symbolic	• The 100-square serves as a symbolic aid to represent numbers in different ways, like patterns with a common difference as well as multiples (Anghileri, 2000, p. 12)
	Equivalent numerical forms (including de-composition and re-composition)	• Dotted and array cards can be used as concrete materials to facilitate multiple representations for numbers (Howden, 1989, p. 7)
	Comparison to benchmarks	• Benchmarking can be encouraged by making use of dot cards and array cards (Muir, 2012, p. 26)

Sense of relative and absolute magnitude of numbers	Comparing to physical referent	• Subitising, which is the instant recognition of the number of items in a group, can be facilitated by making use of dot cards and array cards (Muir, 2012, p. 26)
	Comparing to mathematical referent	• Ten frames can be used to assist in visualising numbers as doubles, neighbours and to look for groups of tens in adding numbers (Howden, 1989, p. 7)
System of benchmarks	Mathematical	• An estimation challenge, with a real-sized picture of the item being estimated printed on a laminated card, can be used to test how learners would apply numerical benchmarks (Muir, 2012)
	Personal	• An estimation challenge, with a real-sized picture of the item being estimated printed on a laminated card, can be used to test how a learner may use his/her personal experience of magnitude to solve a mathematical problem (Muir, 2012)
Understanding the effect of operations	Operating on whole numbers	• Card games can be used to exercise the four operations with whole numbers (Stott, 2014, p. 4). • A "pictorial representation of a number line can be an effective method for modeling the addition of integers" (Cope, 2015, p. 12) • Array and grid representations assist with the multiplication of whole numbers (Barmby et al., 2010)
	Operating on fractions/decimals	• Diagrammatical representations of fractional parts within a whole unit can be used to teach equivalent fractions in order for learners to understand the relationship between the numerators and denominators of equivalent fractions (Askew et al., 1997, p. 30) • The tangram, a Chinese geometrical puzzle consisting of a square cut into seven pieces which can be arranged to make various other shapes, can be used to teach the relationship between fractional parts (Yang et al., 2008, p. 803)
Understanding mathematical properties	Commutativity	• Dotted and array cards can be used as concrete materials to enhance the understanding of mathematical properties (Howden, 1989, p. 7) • Grid representations highlight the commutativity of numbers (Barmby et al., 2010)
	Associativity	• Dotted and array cards can be used as concrete materials to enhance the understanding of mathematical properties (Howden, 1989, p. 7)
	Distributivity	• Dotted and array cards can be used as concrete materials to enhance the understanding of mathematical properties (Howden, 1989, p. 7) • The array representation assists teachers in showing the distributive properties of multiplication (Barmby et al., 2010, p. 48)

(*Continued*)

Number sense understandings	*Number sense themes*	*Visual teaching strategies*
	Identities	• A pictorial representation of a number line can be an effective method for teaching the identity property for addition and subtraction (Cope, 2015) • The array representation can be used to teach the number 'one' as a multiplicative identity (Barmby et al., 2010)
	Inverses	• A pictorial representation of a number line can be an effective method for teaching inverse values (Cope, 2015)
Understanding the relationship between operations	Addition/multiplication	• Use of the array representation helps to "show that multiplication can be the same as repeated addition" (Barmby et al., 2010, p. 47)
	Subtraction/division	• Use of the array representation helps to show that division can be the same as repeated subtraction (Barmby et al., 2010)
	Addition/subtraction	• A pictorial representation of a number line can be an effective method for teaching addition and subtraction as inverse operations (Cope, 2015)
	Multiplication/division	• Dotted and array cards can be used as concrete materials to enhance the understanding of multiplication and division as inverse operations (Howden, 1989)
Understanding the relationship between problem context and the necessary computation	Recognise data as exact or approximate	• An estimation challenge, with a real-sized picture of the item being estimated printed on a laminated card, can be used for allowing learners to explore magnitude as exact or approximate (Muir, 2012)
	Awareness that solutions may be exact or approximate	• An estimation challenge, with a real-sized picture of the item being estimated printed on a laminated card, can be used to create an awareness of the type of solutions needed within the context of the mathematical problem posed (Muir, 2012)
Awareness that multiple strategies exist	Ability to create and/or invent strategies	• An estimation challenge, with a real-sized picture of the item being estimated printed on a laminated card, can be used to test the learner's ability to create or invent solution strategies within the context of the mathematical problem posed (Muir, 2012)
	Ability to apply different strategies	• An estimation challenge, with a real-sized picture of the item being estimated printed on a laminated card, can be used to test the learner's ability to apply different strategies within the context of the mathematical problem posed (Muir, 2012)

	Ability to select an efficient strategy	• An estimation challenge, with a real-sized picture of the item being estimated printed on a laminated card, can be used to test the learner's ability to select an efficient strategy within the context of the mathematical problem posed (Muir, 2012)
Inclination to utilise an efficient representation and/or method	Facility with various methods (mental, calculator, pencil and paper)	• Allowing learners to experiment with unconventional visual algorithms like the lattice method of multiplication and the scaffolding algorithm for division will stimulate their natural ability to use various methods and representations in mathematical problem-solving (Groth, 2012, p. 159)
	Facility choosing efficient number(s)	• An array representation can be used to multiply two awkward numbers. The numbers are rounded up and part of the array is subtracted after the calculation to match the original numbers (Barmby et al., 2010)
Inclination to review data and result for sensibility	Recognise reasonableness of data	• An estimation challenge, with a real-sized picture of the item being estimated printed on a laminated card, can be used to test the learner's ability to recognise the reasonableness of data within the context of the mathematical problem posed (Muir, 2012)
	Recognise reasonableness of calculation	• An estimation challenge, with a real-sized picture of the item being estimated printed on a laminated card, can be used to test the learner's ability to recognise the reasonableness of calculation within the context of the mathematical problem posed (Muir, 2012)

Source: adapted from McIntosh et al., 1992.

The VIP was therefore crafted with a view to facilitate a classroom environment which is conducive to the use of visualisation processes in the context of developing good number sense. Visualisation processes were employed both as a strategy to teach the number sense themes set out in Table 3.3 and as an aid for learner engagement during the lessons.

Social Constructivism

The theoretical aspect which is of integral importance to the learning and development of number sense in this study is social constructivism. This theory of how the learning of mathematics occurs provided a theoretical lens and conceptual framework through which the researcher could investigate the role of visualisation processes as a strategy to enhance the teaching of number sense.

Constructivism is an epistemological construct which grew from an enquiry regarding different ways of knowing. Originally, it was accepted that knowledge was the product of transferred intellectual content from a knowledgeable individual to an uninformed one, until thought leaders transformed this view. The learner was no longer viewed as a mere absorber of knowledge. The constructivist view recognises the learner as an active constructor of knowledge, instead of a passive receiver thereof (Piaget, 1972).

From the work of Piaget and other leading cognitive theorists, a more refined theory of learning was developed. The learning environment is by definition social in nature. As an interplay of human agents informs learning, the theory of social constructivism entered the debate (Vygotsky, 1978). Vygotsky (1978) took the theoretical stance of constructivism a step further by emphasising the social factor involved in the construction of knowledge – hence the idea of social constructivism. He regarded the role of communication, social interaction and instruction as paramount for student learning and development.

This research study reflected on teaching proficiency in the context of social constructivism. As such, it explored the extent to which the participating RUMEP teachers made use of visual teaching strategies for the construction of number sense understandings by their learners, as a result of those teachers participating in the VIP.

Research Methodology

The overall goal of this study was to investigate the role which visualisation played, when employed as a teaching strategy to enhance number sense. The study consisted of two data generating processes. The first one involved selected RUMEP teachers teaching number sense and the second involved administering and analysing a set of pre- and post-tests written by learners.

This study asked the following research question:

How do teachers use visualisation processes in a mathematics classroom in order to enhance their teaching of number sense after participating in the VIP?

This mixed method study was undertaken with a total of 12 Grade 6 teachers who teach mathematics. First, I invited five teachers from well-resourced Northern Cape schools to collaborate with me in designing and crafting the VIP. Second, seven selected RUMEP teachers were then invited to participate in the VIP for a period of three terms. It is the latter group of seven teachers that constituted my case study for the empirical research. I also administered a number sense pre- and post-test to learners of the seven RUMEP teachers, to ascertain whether discernible change in learner performance had taken place as a result of the implementation of the VIP.

Gerring (2004, p. 341) defined a case study as "… an in-depth study of a single unit (a relatively bounded phenomenon) where the scholar's aim is to elucidate features of a larger class of similar phenomena". In this research study, the case was the seven participating teachers teaching number sense themes to Grade 6 mathematics learners. The unit of analysis was the views and responses of the seven participating RUMEP teachers (after participating in the VIP) about the use of visualisation processes. A further unit of analysis was the pre- and post-test results.

Data Collection and Analysis

Data for the qualitative analysis was collected by observing the VIP lessons of the seven participants and interviewing them afterwards. Data for the quantitative analysis was collected from the pre- and post-tests that the learners of the participants wrote.

Data was also collected through 14 individual video-stimulated recall interviews with the participating RUMEP teachers during the reflective workshops after each lesson. Nguyen et al. (2013, p. 2) define video-stimulated recall interviewing as a data collection method which "involves video-taping teachers during their normal teaching duties then playing them the video recordings of their own behaviour".

Pre- and post-tests on number sense were administered to the learners of the seven participating RUMEP teachers, to measure any discernible change in learner performance as a result of the implementation of the VIP. The test consisted of number sense items sourced from the *Number Sense Item Bank* designed by McIntosh et al. (1997). The test was customised for South African learners and questions on four main number sense topics were set.

Analysis of the Data

The research data analysis consisted of the collation, presentation and discussion of the results that were obtained from the focus group interview with the seven participants, the 14 observations of the seven participants' VIP lessons, the 14 video-stimulated recall interviews with the seven participants and the assessment results of the pre- and post-tests. The research data were analysed by looking at the participants' experiences and views on engaging the VIP. The results of the number sense pre- and post-tests which were administered to the learners of the participating teachers were then analysed. Parallel to the analysis of the views and responses of the seven teachers about the use of visualisation processes as a result of participating in the VIP, a narrative was structured across the teachers.

Gall et al. (1996, p. 767) define qualitative research as an:

> inquiry that is grounded in the assumption that individuals construct social reality in the form of meanings and interpretations, and that these constructions tend to be transitory and situational. The dominant methodology is to discover these meanings and interpretations by studying cases intensively in natural settings and subjecting the resulting data to analytic induction.

In analysing the transcripts of the video-stimulated recall interviews, I therefore first looked for concepts which I could group into categories, in order to "facilitate interpretation for meaning" (Hart et al., 2009, p. 29). The categories and subcategories constitute the conceptualisation of the interview data as "codes are abstractions of the data, particularly patterns in the data, not mere summaries" (Glaser, 2010, p. 27).

The concepts which were found in the interview transcripts were grouped, and are described, discussed and interpreted in the following categories:

- Insights of the participant about the use of visualisation processes as a result of participating in the VIP.
- Clarifications of the participant's responses (as a subcategory of insights) about the use of visualisation processes as a result of participating in the VIP.
- Subjective meanings of the participant's responses about the use of visualisation processes as a result of participating in the VIP.
- Perceptions of the participant (as a subcategory of subjective meanings) about the use of visualisation processes as a result of participating in the VIP.
- Errors and misconceptions of the participant about the use of visualisation processes as a result of participating in the VIP.
- Consistencies in the participant's responses about the use of visualisation processes as a result of participating in the VIP.
- Inconsistencies in the participant's responses about the use of visualisation processes as a result of participating in the VIP.
- Common elements of the participant's responses about the use of visualisation processes as a result of participating in the VIP.

The focus group consisted of the researcher (myself), the five members of the collaboration team and the seven participating teachers. The meeting was arranged after the seven participants had completed teaching the pilot lesson of the

VIP and the first semi-structured, video-stimulated recall interview was done. The discussion of the focus group primarily focused on refining the implementation of the VIP and clearing any misconceptions that the seven participants may have had.

The qualitative and quantitative analyses discussed above informed the findings of the research study, which are presented next.

Findings

Findings from the Qualitative Analysis

Teacher 1 was consistent in aligning both lessons with the analytical framework shown in Table 3.4. The only challenge for her was finding a visual strategy for operating on decimals. Teacher 1 was therefore rated favourably regarding her proficiency in using visual strategies in the teaching of number sense, which she learnt in the VIP. Teacher 2 also succeeded in aligning both lessons with the analytical framework in Table 3.4. He chose good visual strategies in general, except for the visual strategy to operate on decimals.

On the other hand, Teacher 3 struggled to align her first official lesson with the analytical framework in Table 3.4 when teaching how to operate on whole numbers and decimals. However, when she taught operations on fractions, she made use of a diagrammatical representation of fractional parts. This visual aid assisted learners with adding and subtracting fractions fluently. Teacher 3's second lesson was consistently aligned with the analytical framework shown in Table 3.4. Her choice of a diagrammatical representation with smaller rectangles within a big rectangle, a decimal number line, a dotted array and the 100-square were relevant for the number sense themes that were evident in her lesson on multiplication and division.

Teacher 4 struggled to align his first lesson with the analytical framework in Table 3.4 when teaching how to operate on decimals. The decimal place value chart was not the best strategy for teaching the addition of decimal values. The teacher did not use any visual strategy to show how to compare different decimal quantities. It would have been more effective to convert decimal fractions to common fractions by making use of a conversion chart and then comparing the quantities.

Teacher 5, Teacher 6 and Teacher 7 struggled to align the teaching of addition and subtraction of decimal fractions with the analytical framework in Table 3.4 during their first lesson. The decimal place value chart which Teacher 5 employed was not the best strategy for teaching the addition of decimal values, while Teacher 6 and Teacher 7 reverted to the conventional way of teaching decimal fractions without seeking innovative visual methods. A series of calibrated decimal number lines could rather have been used for teaching how to operate on decimals, as decimals can be displayed with different calibrations showing ever-decreasing distances between the same start and end point.

The second lessons of all three participants were consistently aligned with the analytical framework shown in Table 3.4. They chose the diagrammatical representation with smaller rectangles within a big rectangle, a decimal number line, the spider diagram, an array of dots and the fraction wall as visual aids for the number sense themes that were evident in their lessons on multiplication and division.

Teacher 5 and Teacher 6 both expressed the view that challenged learners also benefitted from visual teaching. This view was encouraging as it suggested that visualisation processes may facilitate an environment for optimal learning opportunities to ensure that struggling learners gain conceptual understanding.

Findings from the Quantitative Analysis

The pre- and post-test results of the seven experimental schools and the control school were compared, in order to report on the change in learner performance.

Table 3.5 shows the percentage of change in the number sense in the pre- and post-test results of the experimental schools and the control school. All eight schools performed better in the post-test. Five of the seven experimental schools performed better than the control school with regard to the post-tests.

In Table 3.6, the average pre- and post-test results of the experimental schools were calculated and compared to that of the control school. It is clear from the evidence that the control school generally performed better in both the pre- and the post-test. However, most of the experimental schools (five out of seven) demonstrated a better relative improvement than the control school.

Table 3.5 A comparison of the number sense test results in the seven experimental schools and one control school

Schools	*Average Pre-test (%)*	*Average Post-test (%)*	*Average Change (%)*
Experimental school 1	18	25	7
Experimental school 2	21	23	2
Experimental school 3	19	22	3
Experimental school 4	32	39	7
Experimental school 5	27	35	8
Experimental school 6	20	28	8
Experimental school 7	26	38	12
Control school 1	35	40	5

Table 3.6 A comparison of the average number sense test results in the experimental schools and the control school

Schools	*Pre-test (%)*	*Post-test (%)*	*Change (%)*
Experimental schools (average)	23	30	7
Control school	35	40	5

Table 3.7 Average learner performance per number sense topic at the experimental schools

Number sense topics	*Average pre-test (%)*	*Average post-test (%)*
Whole numbers	25	34
Fractions	22	30
Decimals	21	24
Percentages	26	32

Table 3.8 A comparison of learner performance in number sense topics in the control school

Number sense topics	*Average pre-test (%)*	*Average post-test (%)*
Whole numbers	47	56
Fractions	36	39
Decimals	24	26
Percentages	32	41

Table 3.7 shows that the experimental schools improved their performance in all four number sense topics, although both the pre- and post-test results were generally poor. It highlights the fact that decimal fractions are extremely challenging for learners. The results in Table 3.7 indicate that particularly the topics of 'common fractions' and 'whole numbers' improved remarkably due to the use of visual aids.

In Table 3.8, the learner performance of the control school in the different number sense topics was analysed and compared to that of the experimental schools. The experimental schools had an 8% increase in the topic of 'fractions', whereas the control school only improved its performance in the same topic by 3%. The experimental schools also showed more improvement in the topic of 'decimals'. Although the control school performed better in the topic of 'percentages', the overall improved performance of the experimental schools was better than that of the control school.

Conclusion and Recommendations

Significance

The value of this research study should be assessed in terms of the contribution that the use of visualisation processes made to fostering a deeper understanding of number sense themes for the learners. Furthermore, whether learners could "move seamlessly between the real world of quantities and the mathematical world of numbers and numerical expressions" should be determined, to ascertain whether the learners acquired epistemological access to mathematics (Case, 1998, p. 1).

It is interesting to note that despite the Curriculum and Assessment Policy Statement (CAPS) for the Intermediate Phase being introduced in 2013 in South Africa, the poor ANA results in 2013 and 2014 indicated that the expected level of fluency and numeracy was not yet achieved (DBE RSA, 2014). This was perhaps due to challenges with the implementation of the CAPS document in Grade 6. The most important challenges, in my reading and experience, can be summed up as follows:

- Although the CAPS document resonates well with number sense understandings and the analytical framework in Table 3.4, there is no explicit drive in the curriculum statement to advocate the use of visualisation processes. The use of visual aids is simply recommended for some number sense themes in the clarification notes of the curriculum, but no concrete guidelines are provided (DBE RSA, 2011).
- Only 50% of the curriculum is currently allocated for the teaching of the content area related to number sense, namely, numbers, operations and relationships. Because good number sense is a prerequisite for mastering higher order mathematics during later phases, more time should be allocated in the primary mathematics curriculum for the teaching of number sense understandings. It is encouraging that the Mathematics Teaching and Learning Framework of the Department of Education states that issues in the curriculum such as "scope, depth, sequencing and time allocation need to be reviewed" in order to teach mathematics for conceptual understanding (DBE RSA, 2018, p. 10).
- The CAPS document compartmentalises number sense topics. This has led to number sense being taught in silos of the different number sense topics, which makes the effective and coherent development of number sense understandings difficult for learners. Whole numbers and fractional parts should be taught alongside each other using visualisation strategies to expedite the conceptual understanding of the learners.
- Although the CAPS document has constructivist underpinnings, no comprehensive guidance is given to teachers in terms of their classroom practice. The methods and practice of teaching number sense understandings should be infused with the use of visualisation processes to ensure that learners make meaning of the number sense themes as per Table 3.3.

Herein lies the significance of the research study which focused on Grade 6 in the Intermediate Phase in primary schools. The mathematics learning backlog that was apparent in the Intermediate Phase at the primary level is carried forward into the Senior Phase at high schools, as the 2013 ANA results showed that "achievement levels in Mathematics declined across the grades with progressively steeper declines from Grade 6 to Grade 9" (DBE RSA, 2014, p. 15).

There is therefore a need to explore the pedagogics of Intermediate Phase teachers with regard to number sense, in order to fully grasp the nature of the crisis in mathematics teaching and learning, and to contribute to the debate of how best to improve the mathematics performance in South African schools. My research study showed that the use of visualisation processes for the teaching of number sense led to the learners being able to construct their own knowledge by using their own calculating procedures and using the visual aids provided. Kamii et al. (1993) emphasise that learners should be encouraged to invent and use their own calculating procedures. This finding in my research study is a reflection of the extent to which visual teaching enhances the acquisition of number sense.

The development of the participating teachers was enhanced as a result of them taking part in the VIP. In their reflections, the participants expressed their views on how the implementation of the VIP improved the quality of their own classroom practices.

Recommendations

My research study advocates valuable information for teachers to consider when they use visualisation processes in the teaching of number sense. It is important to create a mathematically rich, stimulating and visual environment for the learner groups and guide them towards conceptual understanding of the number sense theme under discussion, then allow them sufficient time for self-exploration in completing the lesson activities by making use of the visual aids.

I further suggest that research be undertaken on the role of visualisation processes in the remedial teaching of Intermediate Phase learners who have difficulties in conceptualising number sense. Lastly, research into an alternative curriculum rooted in visualisation processes is encouraged. This is particularly important in the South African context, where, since the birth of our democracy in 1994, curricula changes have not succeeded in substantially improving the fluency of skills in numeracy of Intermediate Phase learners.

We can therefore rightly argue that South African Intermediate Phase mathematics learners face difficulties regarding epistemological access to mathematics education. On the upside, my research study suggests that epistemological access to Intermediate Phase mathematics education can be improved by the use of visualisation processes in the teaching of number sense understandings as per Table 3.3.

My research study furthermore advocates that enhanced visual teaching of number sense understandings places conceptual understanding first, then subsequently gives attention to procedural fluency. It follows that, in the context of my study, if conceptual understanding is achieved, the learners will understand why the procedure works and they will then be able to deduce the rules for themselves.

References

Alsina, C. & Nelsen, R. (2006). *Math Made Visual: Creating Images for Understanding Mathematics.* https://doi.org/ 10.1017/CBO9781614441007

Anghileri, J. (2000). *Teaching Number Sense.* London: Continuum.

Arcavi, A. (2003). The role of visual representations in the learning of mathematics. *Educational Studies in Mathematics, 52(3),* 215–241. http://dx.doi.org/10.1023/A:1024312321077

Askew, M., Brown, M., Rhodes, V., Wiliam, D. & Johnson, D. (1997). *Effective Teachers of Numeracy: Report of a Study Carried Out for the Teacher Training Agency.* London: King's College, University of London.

Ausubel, D. P. (2000). *The Acquisition and Retention of Knowledge.* Dordrecht: Kluwer.

Barmby, P., Harries, A. V., & Higgins, S. E. (2010). Teaching for understanding/understanding for teaching. *Issues in Teaching Numeracy in Primary Schools,* 45–57. https://www.researchgate.net/publication/221900444_Teaching_for_understandingunderstanding_for_teaching

Boaler, J., Chen, L., Williams, C., & Cordero, M. (2016). Seeing as understanding: The importance of visual mathematics for our brain and learning. *Journal of Applied and Computational Mathematics, 5*(5), 1–17. https://doi.org/10.4172/2168-9679.1000325

Case, R. (1998). *A Psychological Model of Number Sense and its Development.* Paper presented at the annual meeting of the American Educational Research Association, San Diego.

Chrysostomou, M., Pitta-Pantazi, D., Tsingi, C., Cleanthous, E., & Christou, C. (2013). Examining number sense and algebraic reasoning through cognitive styles. *Educational Studies in Mathematics, 83.* https://doi.org/10.1007/s10649-012-9448-0

Cope, L. (2015). Math manipulatives: Making the abstract tangible. *Delta Journal of Education, 5(1),* 10–19.

Department of Basic Education RSA (2011). *Curriculum and Assessment Policy Statement (CAPS): Mathematics Grades 4–6.* Pretoria: DBE.

Department of Basic Education RSA (2014). *Report on the Annual National Assessment of 2014: Grades 1 to 6 and 9.* Pretoria: DBE.

Department of Basic Education RSA (2018). *Mathematics Teaching and Learning Framework for South Africa. Teaching Mathematics for Understanding.* Pretoria: DBE.

Du Plessis, P. & Mestry, R. (2019). Teachers for rural schools – A challenge for South Africa. *South African Journal of Education, 39(1),* 1–9. https://doi.org/10.15700/saje.v39ns1a1774

Gall, M. D., Borg, W. R., & Gall, J. P. (1996). *Educational Research: An introduction.* White Plains, NY: Longman.

Gerring, J. (2004). What is a case study and what is it good for? *The American Political Science Review, 98(2),* 340–354. https://www.jstor.org/stable/4145316

Gersten, R. & Chard, D. (1999). Number sense: Rethinking arithmetic instruction for students with mathematical disabilities. *The Journal of Special Education, 33(1),* 18–28. https://doi/10.1177/002246699903300102

Glaser, B. G. (2010). The future of grounded theory. *The Grounded Theory Review, 9(2),* 1–38. https://doi.org/10.1177/104973299129122199

Groth, R.E. (2012). *Teaching Mathematics in Grades 6-12: Developing Research-Based Instructional Practices.* Washington, DC: SAGE Publications.

Hart, L. C., Smith, S. Z., Swars, S. L., & Smith, M. E. (2009). An examination of research methods in mathematics education (1995–2005). *Journal of Mixed Methods Research, 3(1),* 26–41. https://doi.org/10.1177%2F1558689808325771

Hegarty, M. & Kozhevnikov, M. (1999). Types of visual–spatial representations and mathematical problem solving. *Journal of Educational Psychology, 91(4),* 684–689.

Howden, H. (1989). Teaching number sense. *The Arithmetic Teacher, 36(6),* 6–11.

Kamii, C., Lewis, B., & Livingston, S. (1993). Primary arithmetic: Children inventing their own procedures. *The Arithmetic Teacher, 41(4),* 200–203. https://www.jstor.org/stable/41195981

Kaminski, E. (2002). Promoting mathematical understanding: Number sense in action. *Mathematics Education Research Journal, 14(2),* 133–149. https://doi.org/10.1007/BF03217358

Lotz-Sisitka, H. (2009). Epistemological access as an open question in education. *Journal of Education, 46,* 57–79. http://hdl.handle.net/10962/7123

McIntosh, A., Reys, B. J., & Reys, R. E. (1992). A proposed framework for examining basic number sense. *For the Learning of Mathematics, 12(3),* 2–8. https://www.jstor.org/stable/40248053

McIntosh, A., Reys, B., Reys, R., Bana, J., & Farrell, B. (1997). *Number Sense in School Mathematics: Student Performance in Four Countries.* Perth, Australia: Mathematics, Science and Technology Education Centre, Edith Cowan University.

Moskal, B. M. & Magone, M. E. (2000). Making sense of what students know: Examining the referents, relationships and modes students displayed in response to a decimal task. *Educational Studies in Mathematics, 43,* 313–335. https://doi.org/10.1023/A:1011983602860

Muir, T. (2012). What is a reasonable answer? Ways for students to investigate and develop their number sense. *Australian Primary Mathematics Classroom, 17(1),* 21–28.

Nguyen, N. T., McFadden, A., Tangen, D., & Beutel, D. (2013). Video-stimulated recall interviews in qualitative research. *Paper Presented at the Joint AARE Conference, Adelaide, Australia,* 1–10.

Piaget, J. (1972). *Psychology and Epistemology.* Harmondsworth: Penguin.

Presmeg, N. C. (1986). Visualization in high school mathematics. *For the Learning of Mathematics, 6(3),* 42–46. https://doi.org/10.1007/978-3-030-15789-0_161

SAIDE (2007). *Teaching and Learning Mathematics in Diverse Classrooms.* Adapted from UNISA, materials by: Ingrid Sapire, RADMASTE, University of the Witwatersrand: Draft Pilot Edition, Project coordinated by: Tessa Welch.

Stott, D. (2014). "I've got it!" – A card game for developing number sense and fluency. *Learning and Teaching Mathematics, 17,* 3–6.

Tellis, W. (1997). Introduction to case study. *The Qualitative Report, 3(2),* 1–14. https://doi.org/10.46743/2160–3715/1997.2024

Terao, A., Koedinger, K. R., Sohn, M. H., Qin, Y., Anderson, J. R., & Carter, C. S. (2004). An fMRI study of the interplay of symbolic and visuo-spatial systems in mathematical reasoning. *Proceedings of the 26th Annual Conference of the Cognitive Science Society,* Chicago, IL, 1327–1332.

van Garderen, D. (2006). Spatial visualization, visual imagery, and mathematical problem solving of students with varying abilities. *Journal of Learning Disabilities, 39(6),* 496–506. https://doi.org/10.1177/00222194060390060201

Vygotsky, L. (1978). *Mind in Society.* Cambridge, MA: Harvard University Press.

Yang, D. C., Li, M. N., & Lin, C. J. (2008). A study of the performance of 5th graders in number sense and its relationship to achievement in mathematics. *International Journal of Science and Mathematics Education, 6,* 789–807. https://doi.org/10.1007/s10763-007-9100-0

4 A Case for the Visual Teaching of Fractions

Charity Ausiku

Introduction

The complexity of mathematics as a subject can be attributed to its abstract nature and the emphasis placed on the memorisation of procedures and rules by teachers. Bossé and Bahr (2008, p. 3) assert that "historically, traditional mathematics instruction has been characterized by an extreme commitment to the rote memorization of procedures with little concern for the associated concepts that underlie them." Hence, mathematics is reduced to the mere memorisation and reproduction of rules and facts. Often, the procedures do not make sense to learners and this in turn frustrates many learners and discourages them from pursuing mathematics in advanced grades (Gafoor & Kurukkan, 2015). For mathematics instruction to make sense to learners, it should focus on conceptual understanding. Molina (2014, p. 3) asserts that

> mathematics instruction must first ensure that students' conceptual understanding is deeply embedded. When students have truly mastered a concept, they should be able to show all the detailed steps in a process, explain why those steps occur, and connect the process to related concepts.

Therefore, a paradigm shift in the way mathematics is taught is imperative, since the traditional way of teaching mathematics has proved to be problematic. According to Boaler et al. (2016), visualisation is an alternative approach to mathematics instruction as it provides different ways of seeing, understanding and extending mathematical ideas that have been underdeveloped or missed. Rosken and Rolka (2006, p. 458) further argue that "in mathematics learning, visualisation can be a powerful tool to explore mathematical problems and to give meaning to mathematical concepts and the relationship between them." Although other strategies through which conceptual understanding can be enhanced exist, visualisation was opted for due to the significant impact that it could have on the pedagogy of mathematics in general and fractions in particular.

Generally, fractions have proved to be one of the most difficult concepts to teach and learn in mathematics (Richardson, 2019; Ubah & Bansilal, 2018;

DOI: 10.4324/9781003172420-5

Braithwaite & Siegler, 2018; Van De Walle et al., 2013; Lamberg & Wiest, 2015; Bruce et al., 2013). Moreover, I took a keen interest in my study, the focus of this chapter, after realising that my distance students who were enrolled for the Bachelor of Education (BEd) degree grappled with mathematics concepts, especially fractions. This led to the inception of the Rundu Campus Fraction Project (RCFP), a project that promotes the use of visualisation in teaching fractions by collectively developing visual materials.

The research questions that guided the study are composed of one main research question and four sub-questions, as presented below:

How does the incorporation of visualisation processes in mathematics lessons as a result of teachers' participation in the RCFP enhance the teaching of fractions, if at all?

1 What type of visualisation processes do senior primary school teachers incorporate in their mathematics lessons?
2 How do senior primary school teachers incorporate visualisation processes in their mathematics lessons?
3 What significance do senior primary school teachers attach to the incorporation of visualisation processes in mathematics lessons?
4 What are the enabling and constraining factors in teaching fractions in an explicitly visual way at the senior primary phase?

Dual Coding Theory (DCT) and Constructivism

This study was informed by the DCT and the Constructivist Theory (CT). While the DCT advocates for the recognition of both verbal and non-verbal codes (VNVCs) in teaching and learning (Clark & Paivio, 1991), the CT is founded on the notion that the meaningful acquisition of knowledge involves active construction of knowledge on the part of the learner. Since visualisation is embedded in the non-verbal code, this study aimed to establish the extent to which participants of the RCFP deliberately incorporated visualisation processes in their teaching and whether these processes were meaningfully integrated into the pedagogy of fractions. The RCFP was established to expose teachers to effective fraction instructional strategies encompassing both the VNVCs (Ausiku, 2019). However, since the non-verbal (visual) code is generally not well understood and often neglected, emphasis (during the RCFP training sessions) was placed on the non-verbal code in order to help teachers understand its role in the teaching and learning of fractions.

Consequently, a RCFP manual that focused on basic fraction concepts such as fraction size, comparing fractions, addition, subtraction, multiplication and division of fractions was developed. The RCFP manual is a product of the RCFP and it is used as a reference guide by the participants. It is designed in such a way that each section begins with a brief description of the concept, followed by examples which are presented using standard algorithms and visual methods. This approach to the teaching of fractions is drawn from the DCT.

The use of the two codes in the DCT presents learners with two pathways of learning or recalling information, making it easier for learners to access information easily – regardless of their preferred mode of learning. According to Mayer and Anderson (1991, p. 485),

> this theory predicts that learners will remember and transfer material better if they encode the material both visually and verbally because they have two separate ways of finding the information in memory.

The combined effect of the two codes is believed to be more powerful than the use of either the verbal code or the visual code (Avgerinou & Pettersson, 2011). Focusing on one code (for instance, the verbal code) could deprive the visually oriented learners of opportunities to learn. Therefore, adopting the DCT is one way of making fraction content more accessible to learners in mathematics classrooms.

The Pedagogy and Visualisation of Fractions

Difficulties associated with the pedagogy of fractions can be attributed to the complex nature of fractions. Barnett (2016, p. 12) points out that

> working with fractions is usually the first time in their schooling when students give up on making sense of a concept and resort to simply following the procedure demonstrated by the teacher.

The multiple interpretations of fractions; the unusual way in which fractions are written; learners' overgeneralisation of their whole number knowledge; and, most importantly, the fact that fraction instruction does not focus on the conceptual understanding of fractions are the main reasons why learners struggle with fractions (Van De Walle et al., 2013; Cramer & Whitney, 2010). Subsequently, this calls for effective teaching strategies that can enhance learners' conceptual understanding of fractions rather than the traditional, rule-oriented strategies.

Visualisation is instrumental in making fractions more concrete and comprehensible. Although visualisation has been at the centre of many studies (Kara & Incikabi, 2018; Barnett, 2016; Fennell & Karp, 2017; Bruce et al., 2013; Chen et al., 2013), it has been used in a variety of contexts. This study looked at visualisation from a different angle by focusing on its role as a teaching strategy rather than a teaching aid. It is worth noting that there is a difference between the use of visualisation as a teaching aid and its use as a teaching strategy. Using visualisation as a teaching aid in mathematics renders its role in teaching and learning less important because it is perceived as merely an *aid* rather than another equally important way of representing mathematics concepts. When visualisation becomes an integral component of mathematics pedagogy, its function is elevated and forms part of the way in which we do mathematics.

At the core of this study is the concept of visualisation, a concept that can inspire both teachers and learners "to see mathematics differently, to see the creativity and beauty in mathematics and to understand mathematical ideas" (Boaler et al., 2016, p. 7). Moreover, Chikiwa and Schäfer (2019, p. 1) posit that "visualisation is generally accepted and considered as helpful in mathematics education because of its diverse pedagogic, cognitive and epistemic purposes." A visual approach to the pedagogy of fractions was therefore considered as a possible strategy to mitigate the difficulties encountered by learners. Contrary to the view that visualisation can impede the development of formal mathematical development, this study seeks to determine its complementary role to the development of formal mathematics procedures. According to Arcavi (2003, p. 221), the incorporation of visualisation in mathematics calculations "debugs our intuitions, so that the symbolic solution is not only regarded as correct, but also natural and intuitively convincing." Thus, the combined effect of visualisation and formal mathematics procedures can be considered as intertwined and mutually beneficial to the enhancement of learners' conceptual understanding.

In this study, I argue that visualisation should be portrayed as a way of thinking about and doing fractions, rather than as a mere tool that can be used to aid the understanding of abstract fraction concepts, and shelved or discarded once it has served its purpose. According to Boaler et al. (2016, p. 7), the use of visuals in mathematics classrooms is often perceived as "a prelude to the development of abstract ideas, rather than a tool for seeing and extending mathematical ideas and strengthening important brain networks." When visualisation is presented as a way of doing mathematics – just like symbols and numbers – both learners and teachers may begin to value and adopt it as an alternative and natural way to learn and teach fractions (Boaler et al., 2016). The theoretical framework of this study, which is founded on constructivism, and the DCT and its location in the RCFP are critical aspects that distinguish this study from the other studies.

Furthermore, the way teachers teach has a great impact on the way learners learn. Therefore, this is an important consideration in the way that teachers present fraction content to the learners. Moreover, learners' poor understanding of fractions can adversely affect their understanding of other concepts in mathematics such as algebra, ratio and proportion, and percentages (Bruce et al., 2013). Since the RCFP strives to improve the pedagogy of fractions and content knowledge (CK) of teachers, it was considered as a suitable empirical field for this study. This study further argues that visualisation is a way of reasoning, thinking about and doing mathematics.

Research Methodology

This mixed method study focused on the meaning and interpretation that participants attached to visualisation processes in their teaching. I adopted the interpretive paradigm because it was useful in establishing the meaning attached to the participants' pedagogy in terms of visualisation. Furthermore, I located this study in the interpretive paradigm as it makes provision for multiple realities (Rehman &

Alharthi, 2016). Teachers have different reasons for teaching the way they do, for including or excluding visualisation and for adopting certain teaching strategies.

In order to undertake an in-depth understanding of the phenomenon (Creswell, 2012), in this case, the concept of visualisation, I employed a case study methodology. According to Yin (2004, p. 4), "the case study method is best applied when research addresses descriptive or explanatory questions and aims to produce a first-hand understanding of people and events." In addition to observations, the explanations and interpretations related to visualisation, which were provided by the participants in the questionnaire and interviews, were my key sources of data for this study.

This case study was composed of eight primary school teachers from Kavango East region of Namibia, selected on the basis of their responses to the questionnaire. Although this study was predominantly qualitative (quan→QUAL), quantitative methods were employed in the different stages of stages of data collection. I adopted the sequential-transformative mixed method strategy, which is described as a "two-phase project with a theoretical lens overlaying the sequential procedures" (Creswell, 2009, p. 212). Since this study is founded on the DCT and the CT, these two theories were used as the theoretical lenses through which the participants were selected.

Moreover, this mixed method study is illustrative in design. In other words, it "refers to the use of qualitative data to illustrate quantitative findings, often referred to as putting meat on the bones of dry quantitative findings" (Clark, 2008, p. 263). This entails that the different categories of participants selected for this study were authenticated by the qualitative data collected through observations and interviews. Since I targeted participants who indicated that they use both symbolic and visual strategies to teach fractions, the use of qualitative data collection methods was instrumental in either confirming or refuting the claims in the questionnaire. In addition, the qualitative element also helped to answer the "how" question. For instance, if participants indicated that they value the incorporation of visualisation processes in their lessons, it was deemed necessary to observe how this was done and the extent to which it was done.

For this study, I used a questionnaire, interviews and observations as data collection instruments. The three instruments were structured in such a way that it would help me gauge participants' views on the use of fraction models (FMs), how often they used them, their rationale for using FMs and the value attached to the use of visualisation.

Rundu Campus Fraction Project

This section provides an account of the RCFP, its objectives, participants and activities. The RCFP is a campus-based intervention project that was established in 2017 after due consideration of the difficulties that my Bachelor of Education (BEd) student teachers encountered on the concept of fractions. Full-time and distance students enrolled for the BEd degree in junior and senior primary grappled with a number of mathematics concepts, particularly fractions. This is a topic that proved to be incomprehensible to most of them.

In my attempt to mitigate this problem, I realised that they had a shaky foundation in fractions because they had not mastered the basic facts related to fractions. For instance, most of them could not compare or order fractions in ascending or descending order. Other subtopics such as equivalent fractions, addition, subtraction, multiplication and division of fractions also proved to be challenging to the student teachers. Moreover, the verbal or symbolic explanations which included the application of rules did not make sense to my student teachers. Thus, this compelled me to infuse more visuals in my lessons. The main objective of the RCFP is to equip teachers with Pedagogical Content Knowledge (PCK) and CK using strategies that are informed by the DCT and the constructivist theories.

Participants of the RCFP

The RCFP is composed of secondary and primary school mathematics teachers, student teachers from the University of Namibia (UNAM) Rundu Campus and UNAM Rundu Campus lecturers. They all joined the RCFP on a voluntary basis. The Ministry of Education, through the Regional Mathematics Educational Officer (Mathematics Subject Advisor), is also a stakeholder in this project.

RCFP Activities

Members of the RCFP meet at least twice per semester to discuss different fraction concepts, solve fraction problems using visual and conventional methods and develop visual teaching materials. The RCFP sessions focus on the following subthemes: fraction size, comparing fractions, equivalent fractions, addition, subtraction, multiplication, division of fractions and problem solving. The selection of these subthemes is informed by my experience with the third year full-time and part-time BEd students who struggle with these concepts.

Analysing the Data

In analysing the data collected, I developed a seven-concept framework that was derived from the literature, that is, the DCT, visualisation, fractions and pedagogy. As presented in Table 4.1, there are seven key concepts shaping the data analysis framework, namely, Quality Fraction Knowledge (QFK), FMs, Visuality of Lessons (VL), VNVCs, Quantitative Thinking (QT), PCK and Questioning Strategies (QS).

Findings and Discussion

This section presents an overview of the findings from the three research instruments which were used to elicit the participants' views on the various concepts that were instrumental in answering the research questions of this study. The seven concepts used in analysing the data were derived from literature, informed

Table 4.1 The analytical framework

No	*Pedagogical aspects*	*Definition*	*Description*
1.	QFK	• Achieved when teachers use visuals to enhance "deep and new understandings" of fractions	• Selects relevant visuals that are aligned to the lesson objective(s) • Employs at least two FMs • Able to interchangeably provide explanations of selected fraction concepts using both the visual and the verbal codes
2.	FM	• Different ways of representing fractions, that is, area, length and set models	• Represents unit fractions using the area, length and set models • Encourages the use of area, length and set models in fraction computation
3.	VL	• A person's mathematical visuality is the extent to which that person prefers to use visual methods when attempting mathematical problems which may be solved by both visual and non-visual methods	• The use of more than one manipulative enhances students' understanding and promotes abstraction of the concept from irrelevant perceptual features of a manipulative, such as colour, size or shape
4.	VNVCs	• Refers to the use of both visual and non-visual resources	• A fair integration of both visual and verbal codes in the presentation of fraction lessons and assessment activities
5.	QT	• Refers to the conception of relative size of fractions	• In addition to the verbal representations, the teacher uses visualisation processes to enhance learners' understanding of different fraction concepts
6.	PCK and CK	• PCK refers to the ways of representing and formulating the subject matter that makes it comprehensible to others while CK is the type of knowledge explicitly associated with knowledge of the subject matter (Shulman, 1986, p. 9)	• Tests learners' prior knowledge using visual representations • Enhances learners' understanding of fraction concepts by using both VNVCs • Constantly checks learners' understanding of unit fractions using different FMs
7.	Effective questioning strategies (EQS)	• Refers to the type of questioning that is aimed at developing learning and higher order thinking, promoting imagination, speculation, creative thinking and to pitch a suitable challenge level (Gast, 2009, p. 1)	• Poses questions or assessment activities that encourage creative thought and imaginative or innovative thinking

by the two theories underpinning this study. These encompass the VL, the VNVCs, FMs and PCK. The questionnaire was specifically used to determine the profiles of the participants prior to their selection. This exercise was necessary because I had predetermined criteria for the participants whom I intended to select for this study.

The Visuality of Lessons

This concept was instrumental in addressing the first research question, in which I sought to establish the type of visualisation processes that senior primary school teachers incorporated in their lessons. According to Presmeg (1986, p. 298),

> a person's mathematical visuality is the extent to which that person prefers to use visual methods when attempting mathematical problems which may be solved by both visual and nonvisual methods.

In this context, the VL was used to refer to the incorporation of visuals in the participants' lessons in comparison to the total number of lessons taught. The findings reveal that all the eight participants incorporated visuals in some or all of their fraction lessons. From the onset, the participants indicated that they valued the incorporation of visuals in their lessons. Visuality emerged in all sets of data (i.e. the questionnaire, observations and the interviews) although the extent to which it was emphasised or meaningfully incorporated varied.

Data from the lesson observations showed some discrepancies in the incorporation of visuals, as some lessons were more visual than others and in some cases the use of visuals was completely neglected as the participants relied on the symbolic codes. This was in conformity with participants' responses with regard to the incorporation of visuals in the questionnaire. The inconsistencies observed between the participants' views on the incorporation of visuals and the actual incorporation of the visuals can be attributed to a number of pedagogical factors.

Among the factors cited by the participants (in the interviews) as contributing to learners' poor performance in fractions, pedagogical issues came out strongly as major contributing factors. Moreover, the participants also cited the lack of teaching materials, overcrowded classrooms and time allocation as contributing to learners' poor performance in fractions. These factors can be ascribed to the inconsistencies observed in the use of visuals, as some of the participants were more concerned about completing the syllabus than enhancing learners' conceptual understanding. In addition, the lack of ready-made visual materials and the predominantly symbolic assessment methods were mentioned as contributing factors to the way fractions are taught.

In summary, inconsistencies on the VL were observed from all the data sets, that is, the questionnaire, observations and interviews. In other words, this was an indication that the participants used both the verbal and the non-verbal codes. Table 4.2 provides a summary of the visuality of the lessons observed.

Table 4.2 The overall use of visuals in the lessons observed

No	*Participant*	*Incorporation of visuals*	*%*
1	Teacher D	Three out of four lessons $\left(\frac{3}{4}\right)$	75
2	Teacher K	Two out of three lessons $\left(\frac{2}{3}\right)$	67
3	Teacher M1	Two out of four lessons $\left(\frac{2}{4}\right)$	50
4	Teacher S1	One out of three lessons $\left(\frac{1}{3}\right)$	33
5	Teacher M2	One out of three lessons $\left(\frac{1}{3}\right)$	33
6	Teacher H	Three out of four lessons $\left(\frac{3}{4}\right)$	75
7	Teacher I	One out of two lessons $\left(\frac{1}{2}\right)$	50
8	Teacher S2	All three lessons $\left(\frac{3}{3}\right)$	100
			Avg: 60.4%

On average, visuals were incorporated in 60% of the lessons presented. However, in order to fully address the first research sub-question, it was necessary to consider the types of visuals that were used by the participants. Thus, the next section provides an overview of the types of visuals used by the participants, focusing on FMs.

Fraction Models

FMs are important in the teaching and learning of fractions because they present learners with multiple ways to learn fractions. Therefore, this is an important aspect to consider in the pedagogy of fractions. Based on the findings from the questionnaire, some participants opted for the use of one FM while others preferred to use two or three FMs. At this stage of the study, the participants were not asked to justify their choices since these sets of questions were merely used to identify suitable participants for this study. The participants' responses are presented in Table 4.3.

The participants used different types of FMs depending on the topic. In some instances, there was a mismatch between the FMs that the participants said they

Table 4.3 Participants' preferred models

Participants	*Preferred models*		
	Area	*Length*	*Set*
Teacher D	x		
Teacher K			x
Teacher M1	x	x	
Teacher S1	x		
Teacher M2	x		
Teacher H	x		
Teacher I	x	x	
Teacher S2	x		

would use (in the questionnaire) and what they actually ended up using in their classrooms. For instance, in the questionnaire, Teacher D indicated that he preferred to use the area and number line models and this was observed in his lessons. In another example, Teacher S1 indicated that she preferred to use the length model, while she barely used any of the three models in her lessons.

The presence of multiple FMs in the participants' lessons can be attributed to two factors: (1) exposure to visualisation activities in the RCFP and (2) the availability of a reference manual (RCFP manual). On the contrary, the inconsistences observed between the participants' responses to the question about the use of FMs in the questionnaire and the FMs seen during the lesson observations can be attributed to the lack of context in the questionnaire. In the questionnaire, the participants were asked to choose the types of models that they would use to teach fractions in general, without referring to specific topics or fraction concepts. Generally, the topic or fraction concept taught had a major impact on the incorporation or omission of FMs. For example, topics such as the addition and subtraction of fractions, comparing and ordering fractions, fractional parts of a quantity, multiplication of fractions and an assessment activity are some of the topics in which the use of visuals or FMs was minimal or lacking.

Among the three categories of FMs, the area model was more prevalent in all three sets of data, that is, data from the questionnaire, lesson observations and interviews. Apart from Teacher D who made reference to the fraction wall (which falls under the length model) the remaining participants cited examples that fall into the category of area models. In some instances, the examples mentioned by the participants such as diagrams and posters were ambiguous, hence they could be placed in any of the three categories of FMs. The dominance of the area model affirms Cramer and Whitney's (2010) views regarding the overuse of the area model.

Verbal and Non-Verbal Codes

This concept was drawn from one of the theories underpinning this study, that is, the DCT – which is a major element of this study. The main purpose of this study was to determine whether the incorporation of visualisation processes (non-verbal) in the predominantly symbolic (verbal) pedagogy of fractions would enhance the participants' pedagogy of fractions. Mayer and Anderson (1991) argue that the use of both of the VNVCs is vital because it presents learners with two separate pathways to retrieve content from memory. Hence, the VNVC was a key element in the selection of participants because my intention from the onset was to identify participants whose pedagogy was informed by the principles of the DCT.

Notably, some of the participants pointed out that the use of visuals is appropriate for some fraction concepts but not all. By implication, this meant that some fraction concepts are best taught without visuals. Discrepancies were also observed in the levels or phases that the participants considered to be appropriate for the incorporation of visuals. While half of the participants supported the use of visualisation at all levels of education, others felt that it was only appropriate for specific phases such as junior primary and senior primary.

The findings from the observed lessons indicate that in most instances the incorporation of visuals had a direct impact on the Quality of Fraction Knowledge (QFK) portrayed by the participants. The appropriate selection and incorporation of visuals yielded positive results in terms of the pedagogy. As observed in some of the participants' lessons, the careful selection and incorporation of visuals was a major determining factor in the success of the non-verbal code (visual). In some cases, the incorporation of visuals had no impact on the teaching trajectory because the use of these visuals was not clearly aligned with the lesson objectives. In other words, there was a missing link between the non-verbal code, the verbal code and the lesson objectives.

Furthermore, findings from the lesson observations also revealed that a fair understanding of both the verbal and the non-verbal language is essential in teaching fractions effectively. From some of the observed lessons, it emerged that some of the participants experienced language problems related to both the verbal and the non-verbal codes. In terms of the verbal code, a common challenge among the participants was the naming of fractions because fractions such as $\frac{5}{6}$ were referred to as 'five over six' or 'five out of six' rather than 'five sixths.' According to Van De Walle et al. (2013), incorrect fraction language hinders learners' conceptual understanding of the concept of a fraction, as they begin to perceive the two numbers that make a fraction as two separate and unrelated numbers. The use of correct language such as 'five sixths,' combined with the use of visuals, promotes the understanding of a fraction as a single value that is made up of equal-sized parts (Van De Walle et al., 2013; Cramer & Whitney, 2010).

In some instances, this challenge seemed insignificant and had little impact on the pedagogy of participants; however, in fraction lessons that involved word problems, this was revealed as a major challenge. For example, in her lesson on word problems, Teacher M2 read "five-sixths of a bucket" as "*five out of six of a bucket*" – which can easily be misinterpreted by the learners. Consider the two ways in which the following fraction problem was phrased: (1) What is three quarters of eight? (2) What is three out of four of eight? The first question portrays the 'five' and the 'six' in the five sixths as one number rather than two separate numbers, while the second question portrays the 'five' and the 'six' as two separate numbers which may be confusing to learners.

The use of both the VNVCs is beneficial when the two codes are appropriately integrated. However, it is important to note that matching mathematical symbols with visuals was challenging and this can be attributed to the infrequent use of the non-verbal codes by teachers and the complex nature of fraction concepts. Thus, participants were seen grappling with both the VNVCs in terms of language.

PCK was identified as a crucial aspect in this study because of its direct impact on the teaching methods made by the participants. The foundational nature of PCK in relation to the other components of the analytical framework is a vital aspect to consider in the pedagogy of fractions, because it has a direct impact on the success or failure of a lesson. For instance, the decisions to incorporate visuals, to use a variety of FMs, to address learners' misconceptions, to ask relevant questions and match mathematical symbols with visuals were all influenced by the participants' PCK. According to Shulman (1986, p. 9), PCK refers to the ways of "representing and formulating the subject matter that makes it comprehensible to others." The discrepancies observed in the way the participants presented their lessons can be attributed to the differences in their PCK. Furthermore, Star and Stylianides (2013) concur that the level of the teachers' PCK affects their disposition towards teaching and the instructional decisions that they make.

For this study, there was adequate evidence to suggest that the participants who had good PCK incorporated visualisation processes fairly well, as they selected relevant visuals that were appropriately aligned with their lesson objectives. Moreover, these participants possessed effective questioning skills which provided contexts for the integration of the VNVCs. Due to the complex nature of fractions, the contextualisation of concepts such as fraction multiplication is vital. Hence, the need for effective PCK.

In addition, CK and PCK proved to be an indispensable aspect in enhancing the effective pedagogy of fractions. The lack of CK among the participants was revealed through their inability to identify common multiples, lowest/least common denominators, highest common factors and, in some instances, the presentation of incorrect content. Notably, there was a tendency among those who did not have adequate subject matter knowledge to cling to rules and procedures. Fazio and Siegler (2011) assert that teachers' solid conceptual understanding of fractions and their knowledge of learners' common misconceptions are important in improving the quality and pedagogy of fraction lessons. Moreover, participants who lacked CK were observed using long, cumbersome strategies such as the listing of multiples just to find the common multiples or factors of numbers involved in fraction computations. In some instances, the participants were uncertain about the answers provided by their learners.

Conclusion

Based on the findings presented in this chapter, the research questions for this study were all addressed. First, the overarching research goal was addressed in the sense that there was enough evidence to suggest that the RCFP activities had an impact on the pedagogy of the participants. Visualisation processes were effectively incorporated in most of the lessons and this enriched the PCK of the participants. In other words, through the incorporation of visualisation processes, the participants presented learners with more opportunities to learn fractions.

Second, the first sub-question of my research which sought to determine the types of visualisation processes used by senior primary school teachers was also addressed through the use of different FMs. Despite the dominance of the area model, the length and number line models were also prevalent in the participants' lessons and this had a great impact on their pedagogy. In most instances, they relied on drawing their own circles, rectangles, number lines and sets of objects to explain fraction concepts. Moreover, the use of chalkboard visuals allowed learners to experience the procedures involved in the use of visual processes. Some of the participants also drew on the examples presented in the RCFP manual.

Third, there was enough evidence from the lesson observations to answer the third question about how senior primary school teachers incorporate visualisation strategies. The participants partitioned area, length and set models into equal parts to teach different fraction concepts such as fraction size, comparing fractions, adding, subtracting, multiplying and dividing fractions. In addition, concrete objects such as oranges were also used to demonstrate fraction size. One of the challenges related to this aspect was the little attention that participants paid to equi-partitioning.

Fourth, the question about the enabling and constraining factors regarding the explicit integration of visualisation was addressed in my findings and discussion about the VL, FMs, VNVCs and PCK. The enabling factors include: (1) the incorporation of more visuals to improve the participants' pedagogy; (2) the alignment of visuals with the lesson objectives; (3) the need to find a link between the non-verbal and the verbal codes; and (4) the need for adequate CK and PCK. However, several constraining factors were also identified and these include: (1) lack of consistency in the use of visuals; (2) the dominancy of one type of model, that is, the area model; (3) lack of CK and PCK; (4) an imbalance in the use of verbal and the non-verbal codes; and (5) poor synchronisation of fraction language and visual models.

In short, although the incorporation of visualisation processes can improve the pedagogy of fractions, careful consideration about *what, when, how* and *why* visuals should be incorporated is very important, because the inappropriate use of visuals can have detrimental effects on the pedagogy of fractions. However, the opposite is true if visuals are appropriately merged in the teaching of fractions. Therefore, the use of visuals should be encouraged because, if properly incorporated, they have the potential to produce desirable results in the pedagogy of fractions.

References

Arcavi, A. (2003). The role of visual representations in the learning of mathematics. *Educational Studies in Mathematics, 52*(3), 215–241.

Ausiku, C. (2019). Rundu Campus Fraction Project Manual. Unpublished manuscript.

Avgerinou, M. D., & Pettersson, R. (2011). Toward a cohesive theory of visual literacy. *Journal of Visual Literacy, 30*(2), 1–19.

Barnett, J. E. (2016). Transitioning students from the area model to the number line model when developing fraction comparison strategies. MSU Graduate thesis, Missouri State University, Springfield, MO.

Boaler, J., Chen, L., Williams, C., & Cordero, M. (2016). Seeing as understanding: The importance of visual mathematics for our brain and learning. *Journal of Applied & Computational Mathematics, 5*(5), 1–6.

Bossé, M. J., & Bahr, D. L. (2008). The state of balance between procedural knowledge and conceptual understanding in mathematics teacher education. *International Journal of Mathematics Teaching and Learning, 25*(11), 1–28.

Braithwaite, D. W., & Siegler, R. S. (2018). Developmental changes in the whole number bias. *Developmental Science, 21*(2), e12541. http://dx.doi.org/10.1111/desc.12541

Bruce, C., Chang, D., Flynn, T., & Yearley, S. (2013). *Foundations to learning and teaching fractions: Addition and subtraction.* Retrieved July, 4, 2014.

Chen, C. H., Chiu, C. H., Lin, C. P., Wu, S. T., & Hung, Y. C. (2013). Presenting solution strategies of fraction multiplication and division on mathematics instructional websites. *World Journal on Educational Technology, 5*(3), 431–444.

Chikiwa, C., & Schäfer, M. (2019). Visualisation processes in selected Namibian mathematics classrooms. In J. Kriek et al. (Eds.). *Towards Effective Teaching and Meaningful Learning in Mathematics, Science and Technology Education.* 10th ISTE 2019 International Conference, Kruger National Park, Mopani camp, Phalaborwa, Limpopo, South Africa; from 20–25 October 2019.

Clark, J. M., & Paivio, A. (1991). Dual coding theory and education. *Educational Psychology Review, 3*(3), 149–210.

Clark, V. L. P. (2008). *The Mixed Methods Reader.* Thousand Oaks, CA: Sage.

Cramer, K. A., & Whitney, S. (2010). Learning rational number concepts and skills in elementary classrooms: Translating research to the elementary classroom. In *Teaching and Learning Mathematics: Translating Research to the Elementary Classroom,* (pp. 15–22). Reston, VA: NCTM.

Creswell, J. W. (2009). Mixed methods procedures. *Research Design: Qualitative, Quantitative, and Mixed Methods Approaches, 3,* 203–225.

Creswell, J. W. (2012). *Educational Research: Planning, Conducting and Evaluating Quantitative and Qualitative Research* (4th Ed.). Boston, MA: Pearson.

Fazio, L., & Siegler, R. (2011). Teaching Fractions. Educational Practices Series-22. *UNESCO International Bureau of Education.*

Fennell, F., & Karp, K. (2017). Fraction sense: Foundational understandings. *Journal of Learning Disabilities, 50*(6), 648–650.

Gafoor, K. A., & Kurukkan, A. (2015). *Why High School Students Feel Mathematics Difficult? An Exploration of Affective Beliefs.* Paper presented in UGC Sponsored National Seminar on Pedagogy of Teacher Education- Trends and Challenges at Farook Training College, Kozhikode, Kerala on 18th and 19th August 2015. *Online Submission.*

Gast, G. (2009). *Effective Questioning and Classroom Talk: To Develop Learning & Higher Order Thinking, Promoting Imagination, Speculation, Creative Thinking & To Pitch a*

Suitable Challenge Level. Retrieved from http://www.nsead.org/downloads/Effective_Questioning&Talk.pdf

Kara, F., & Incikabi, L. (2018). Sixth grade students' skills of using multiple representations in addition and subtraction operations in fractions. *International Electronic Journal of Elementary Education, 10*(4), 463–474.

Lamberg, T., & Wiest, L. R. (2015). Dividing fractions using an area model: A look at in-service teachers' understanding. *Mathematics Teacher Education and Development, 17*(1), 30–43.

Mayer, R. E., & Anderson, R. B. (1991). Animations need narrations: An experimental test of a dual-coding hypothesis. *Journal of Educational Psychology, 83*(4), 484.

Molina, C. (2014). Teaching mathematics conceptually. *SEDL Insights, 1*(4). SEDL.

Presmeg, N. C. (1986). Visualisation and mathematical giftedness. *Educational Studies in Mathematics, 17*(3), 297–311.

Rehman, A. A., & Alharthi, K. (2016). An introduction to research paradigms. *International Journal of Educational Investigations, 3*(8), 51–59.

Richardson, S. (2019). the relationship between teaching, learning and digital assessment: Final Report. *International Baccalaureate Organization.* https://research.acer.edu.au/ar_misc/37

Rosken, B., & Rolka. K. (2006). A picture is worth a 1000 Words. The role of visualization in mathematics learning. *Proceeding of the 30th Conference of The International Group for The Psychology of Mathematics Education, 4*, 457–464. Prague: PME.

Shulman, L. S. (1986). Those who understand: Knowledge growth in teaching. *Educational Researcher, 15*(2), 4–14.

Star, J. R., & Stylianides, G. J. (2013). Procedural and conceptual knowledge: Exploring the gap between knowledge type and knowledge quality. *Canadian Journal of Science, Mathematics and Technology Education, 13*(2), 169–181.

Ubah, I. J. A. & Bansilal, S. (2018). Pre-service primary mathematics teachers' understanding of fractions: An action-process-object-schema perspective. *South African Journal of Childhood Education, 8*(2), a539. https://doi.org/10.4102/sajce.v8i2.539

Van de Walle, J. A., Karp, K. S., & Bay-Williams, J. M. (2013). *Elementary and Middle School Mathematics: Teaching Developmentally* (8th Ed.). Boston, MA: Pearson.

Yin, R. K. (2004). *The Case Study Anthology.* Thousand Oaks, CA: Sage.

Part 2

Visualisation and Learning

5 Figural Pattern Generalisation

More than Meets the Eye

Duncan Samson

Introduction

The generalisation of number patterns is a widely used didactic approach to engaging with introductory algebra as well as promoting algebraic reasoning. Presenting such number patterns in a pictorial context allows for a far greater depth of engagement compared with patterns presented purely numerically. While numeric patterns have the danger of becoming superficial 'pattern spotting' tasks which can be generalised using basic algorithmic methods, number patterns that are presented pictorially provide a rich context for multiple solution strategies. This has the significant benefit of allowing pupils to develop a variety of equivalent expressions of generality, each of which is visually rooted in the pictorial context. Such an approach has the potential to lead to genuine mathematical exploration, to allow pupils to engage with each other's 'ways of seeing', and thus become appreciative of, and sensitised to, different forms of mathematical sense-making.

Number patterns represented in a pictorial context are thus far more than simply a visual analogue of a numeric pattern. In principle then, a pictorial approach to pattern generalisation has the potential to afford a rich and meaningful context in which to explore notions of generality. However, to what extent are pupils able to generalise pictorial patterns in multiple ways using a variety of visually mediated approaches? An exploration of precisely this question revealed a surprising and unexpected richness in terms of pupils' visualisations. It is the purpose of this chapter to share some of these experiences and to synthesise them into possible pedagogical strategies that could be used to encourage different 'ways of seeing'.

Theoretical Background

Visually mediated approaches to pattern generalisation in a pictorial context provide for an interesting interplay between two different modes of visual perception – *sensory* and *cognitive*. These different modes of perception resonate with Fischbein's (1993) theory of figural objects, and the notion that all geometrical figures possess, simultaneously, both conceptual and figural

DOI: 10.4324/9781003172420-7

properties. Duval (1995) distinguishes between four different modes of figural apprehension – ***perceptual***, ***sequential***, ***discursive*** and ***operative***. Although Duval originally described these modes of apprehension with reference to a classical geometry context, their modification to the context of pictorial pattern generalisation proved a useful way of teasing out the complex interplay between the sensory (figural) and cognitive (conceptual) modes of visual perception.

- **Perceptual apprehension** refers to the *initial* apprehension of a figure, in other words that which is perceived at first glance through the unconscious integration of the gestalt laws of figural organisation.
- **Sequential apprehension** relates to the emergence of sub-figures or elementary figural units stemming from the physical or conceptual construction of the perceived figure.
- **Discursive apprehension** is a process of perceptual recognition during which certain gestalt configurations gain prominence due to an association with discursive statements that may have accompanied the figure.
- **Operative apprehension** relates to the various ways by which a given figure can be modified, through a reconfiguration of the whole-part relation, while retaining its physical integrity.

  Within the context of pictorial pattern generalisation, these different modes of apprehension have the potential to evoke different (yet algebraically equivalent) expressions of generality. Examples of these four modes of figural apprehension, along with their associated expressions of generality, are illustrated in Figure 5.2 for Term 3 of the pictorial sequence shown in Figure 5.1.
- **Perceptual apprehension** could lead to the visual stimulus being seen in terms of squares and triangles as a result of the gestalt laws of closure and good form. This could then potentially lead to the general expression $T_n = 4n + 3n - (n-1) - n$, which takes into account and corrects for overlapping matchsticks.
- **Sequential apprehension** could arise by virtue of the visual stimulus being part of a consecutive sequence of terms. A physical or mental construction of a term from the preceding one in the sequence could potentially foreground the five-match additive unit, which could lead to the general expression $T_n = 6 + 5(n-1)$.

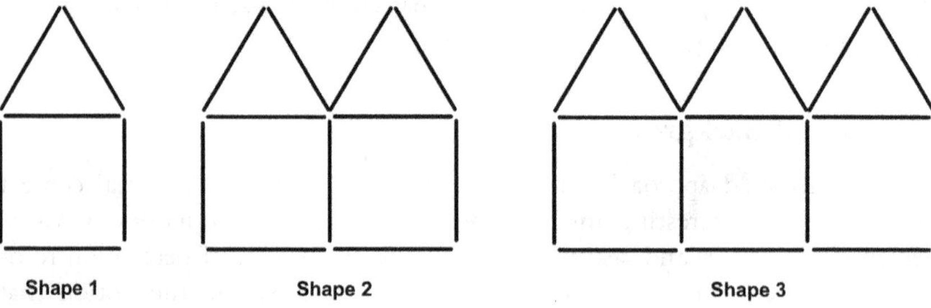

Figure 5.1 The first three terms of a typical pictorial pattern made from matchsticks.

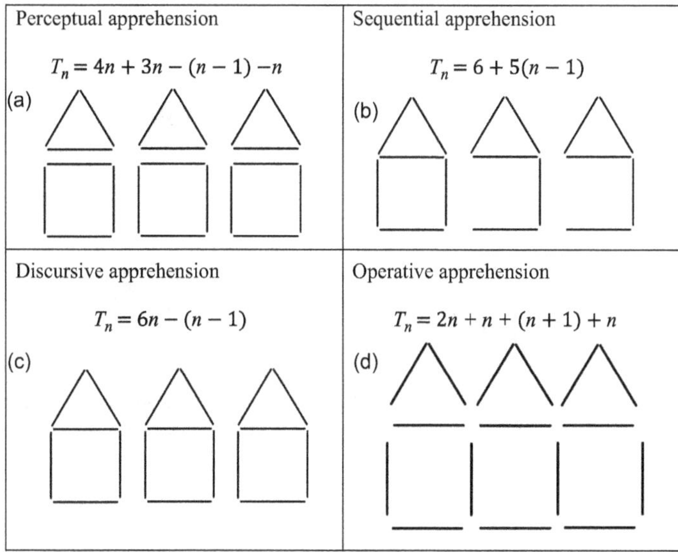

Figure 5.2 Different modes of figural apprehension.

- **Discursive apprehension** could be invoked, for example, by accompanying the visual stimulus with wording incorporating the word *houses* – e.g. "for three houses you need sixteen matchsticks". This could potentially foreground the structural unit of a 'house', leading to the general expression $T_n = 6n - (n - 1)$.
- **Operative apprehension** could bring about a reconfiguration of the whole-part relation allowing the visual stimulus to be seen in terms of a row of inverted V-shapes at the top, a row of vertical matchsticks and two separate rows of horizontal matchsticks. This particular apprehension could lead to the general expression $T_n = 2n + n + (n + 1) + n$.

As illustrated in Figure 5.2, pictorial patterns generally contain a great variety of constituent gestalts and sub-configurations, and it is this multiplicity that represents the *heuristic power* of the pictorial setting. Within the context of figural pattern generalisation, the processes of visualisation and generalisation are deeply interwoven. There is a *phenomenological* element related to the grasping of a generality from the visual stimulus and the different apprehensions that may be evoked, and a *semiotic* element related to the articulation of this perceived generality. Radford's construct of *knowledge objectification* (Radford, 2006, 2008) is a useful lens through which to explore pupils' engagement with generalisation activities presented in a pictorial context. Objectification in this sense can be thought of as a multi-systemic, semiotic-mediated process through which the perceptual act of noticing progressively unfolds – a process of "concept-noticing and sense-making"

(Radford, 2006, p. 15). This represents a fully embodied notion of cognition in which semiotic means of objectification such as spoken or written words, linguistic devices, gestures, rhythm, drawings, signs and the use of artefacts are not simply epiphenomena, but are seen to play a fundamental role en route to a stable form of awareness. As Rowlands (2006) notes, the notion of exploration is one of activity rather than passivity, "it is something we do, rather than something that happens to us" (p. 12). Whole-bodied exploration of a given pictorial context is thus a crucial aspect of the generalisation process. Importantly, such a multi-semiotic view takes cognisance of the notion that different semiotic systems allow for different forms of expressivity and hence play different roles in the objectification process. Radford's construct of ***knowledge objectification*** foregrounds the phenomenological and semiotic aspects of figural pattern generalisation and allows one to critically engage with pupils' whole-body experience and expression as they explore the potentialities afforded by a given pictorial context. At the same time, a modified form of Duval's concept of figural apprehension provides a meaningful way of discussing visual aspects of the phenomenological realm.

Methodology and Data Generation

A mixed gender, high-ability class of 23 Grade 9 pupils at an independent school in South Africa formed the research participants for this study. Since the data collection protocol required pupils to provide verbal articulations of their reasoning process, a high-ability group was deemed more suitable to this particular methodology and more likely to elicit rich data. Research participants were individually provided with two non-consecutive terms of a linear pictorial sequence and were asked, in the space of approximately one hour, to provide as many different expressions for the *n*th term as they could, along with justifications/explanations of their various expressions. Tools such as paper, pencils and highlighters, as well as appropriate manipulatives such as matchsticks and plastic counters, were provided. The participants were asked to think aloud while engaged with their particular pattern generalisation task, and the researcher also prompted the participants to keep talking or provide further explication as and when necessary. Each session was audio-visually recorded, and field notes were also taken by the researcher. In essence, the study aimed to gain insights into the embodied processes of pupils' visualisation activity while engaged in figural pattern generalisation tasks, through an in-depth analysis of each pupil's lived experience. The audio-visual recordings were thus analysed with specific reference to participants' use of semiotic means of objectification such as words, linguistic devices, metaphor, gestures, rhythm, graphics and physical artefacts.

Evidence of Multiple Visual Apprehensions

A broad overview of the data revealed a remarkable ability of some pupils to generalise pictorial patterns through a variety of visual apprehensions. By way of illustration, Table 5.1 shows the nine generalisations arrived at by Grant for the given visual stimulus. The table shows the various expressions for T_n as arrived at by the participant, along with the associated visual apprehension as evidenced by the participant's justification and semiotic means of objectification.

Table 5.1 Grant's nine expressions for T_n

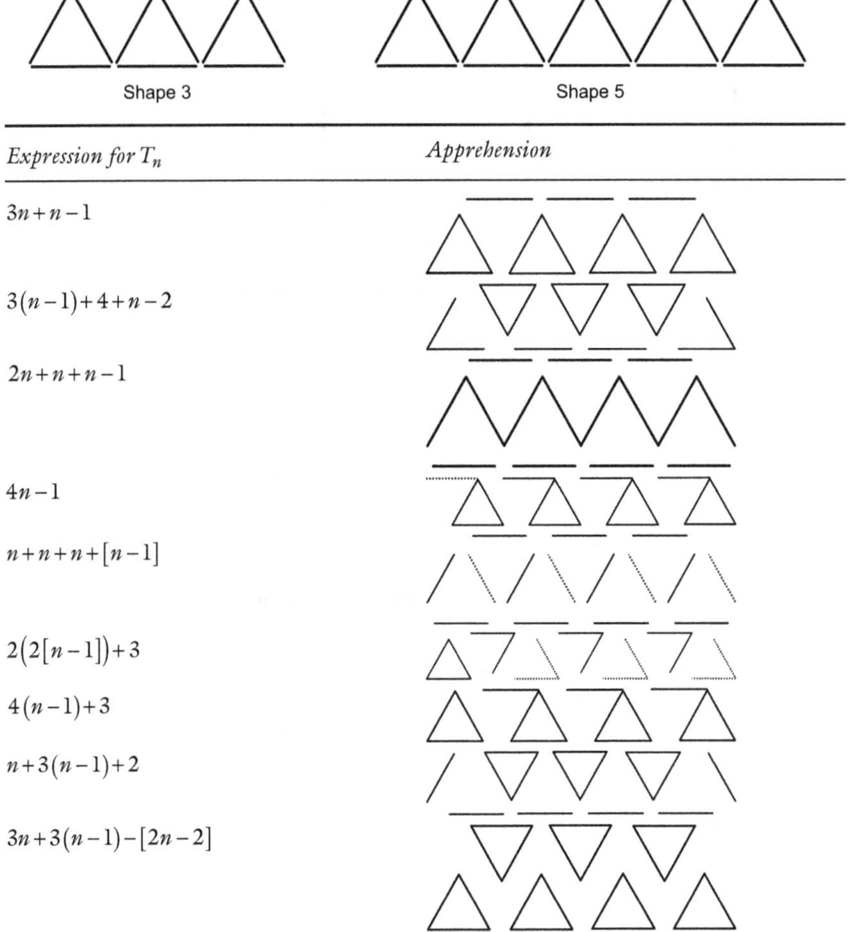

Expression for T_n	*Apprehension*
$3n+n-1$	
$3(n-1)+4+n-2$	
$2n+n+n-1$	
$4n-1$	
$n+n+n+[n-1]$	
$2(2[n-1])+3$	
$4(n-1)+3$	
$n+3(n-1)+2$	
$3n+3(n-1)-[2n-2]$	

Visualisation and Generalisation Strategies

While some pupils favoured a numerical approach to the generalisation activities, those who preferred a visual approach were able not only to move between different apprehensions of the visual stimulus, but also to establish and articulate the generality of their various apprehensions. From a micro-analysis of the audio-visual recordings of pupils whose generalisation approaches were particularly visually mediated (Anthea, Brian, Grant, Kelly, Lance, Liza, Philip and

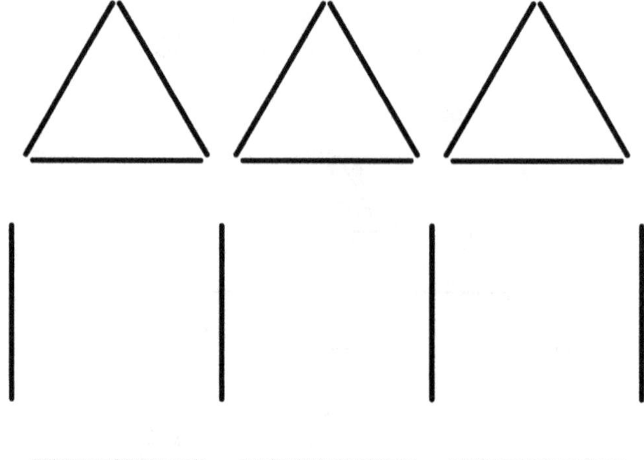

Figure 5.3 A visual reconfiguration centred on the $(n+1)$ vertical matchsticks.

Terry), a rich tapestry of embodied visualisation and generalisation activity was revealed. These processes are now synthesised into possible pedagogical strategies that could be used to encourage different 'ways of seeing'.

Strategy 1: Conscious Visual Engagement

A number of research participants were not only consciously aware of how they were engaging with the pictorial patterns in a general sense, but they were also able to articulate this general strategy. Brian's broad approach was to identify structural 'features' that occurred as many times as the term number itself, or which contained the same number of individual elements as the term number. If, for example, he was looking at Term 5 of a sequence, his conscious strategy would be to look for substructures that occurred five times (e.g. five squares) or features that contained five elements (e.g. a row of five vertical or horizontal matchsticks). In his own words, "*It's sort of like into my head that I must look at it to do with the five* [i.e. the term number] *straight away*". With reference to Figure 5.1, one could, for example, use this strategy to identify that each term contains n squares, or that each term contains two horizontal rows of n matchsticks. One could apply this strategy more generally by consciously seeking out substructures that occur *almost* as many times as the term number, or which contain *almost* the same number of individual elements as the term number. Using this strategy with Figure 5.1 might lead one to focus on the $(n+1)$ vertical matchsticks. This in turn could potentially lead to a visual transformation in which the whole is seen in terms of vertical matchsticks in the middle, a row of horizontal matchsticks at the bottom and a row of triangles at the top (Figure 5.3), leading to the general term $T_n = (n+1) + n + 3n$. Lance's approach, although less overtly articulated than Brian's, was precisely this more general strategy – i.e. to subdivide the pictorial terms into substructures containing either exactly or *nearly as many* elements as the term number itself.

Philip's broad strategy was to identify any striking visual features of the pictorial terms and to use those particular features as the basis of his apprehensions. In his own words, "*I just picked a feature and just tried to work from that*"; "*I just picked a feature again*"; "*Find another feature!*" Sometimes this feature was a recurring element in the diagram, while on other occasions it was a solitary feature that served as a keystone for the rest of his visual apprehension. Terry made use of a similar strategy to that described by Philip – "*...like I'll see a particular shape that stands out, and use that as almost a basis point, like I used the triangles, it stood out a bit, then used the lines, used the squares...*" Terry's comment highlights the importance of being able to move between different visual apprehensions. Consciously seeking out different structural features of the pictorial terms is a way of guarding against a specific apprehension leading to such striking visual imagery that it invokes perceptual rigidity or inflexibility, thereby obscuring other potentially useful gestalts. In addition, pupils should be wary of focusing only on a single term. Within the context of pattern generalisation, the crucial aim is that of evoking a sense of generality. Extended focus on a single term may thus be counter-productive, and pupils should be encouraged to seek out commonalities by moving between different terms in the pictorial sequence.

Strategy 2: Capitalising on Symbolic Ambiguity

An interesting observation that emerged from a micro-analysis of some of the audio-visual recordings relates to the inherent ambiguity in expressions such as $3n$ or $2(n-1)$. By way of example, Philip was presented with the same two pictorial terms as Grant (illustrated at the top of Table 5.1). Of the various apprehensions that he evoked, the third and sixth – both of which he expressed symbolically as $n+2n+(n-1)$ – were based on subtly different visual apprehensions. These are illustrated in Figure 5.4 using Term 5 as a generic example. While both visualisations see the pictorial terms as comprising a horizontal row of n matchsticks along the bottom and a horizontal row of $(n-1)$ matchsticks at the top, there is a subtle distinction in the interpretation of the $2n$ of his general expression. In Figure 5.4a the $2n$ represents two multiples of n matchsticks (the two zigzag sections of his apprehension) while in Figure 5.4b the $2n$ represents n multiples of two matchsticks (the series of inverted V-shapes).

This is an important consideration to bear in mind in conjunction with strategies like that expressed by Brian: "*It's sort of like into my head that I must look at it to do with the five* [i.e. the term number] *straight away*". There are two distinct 'ways of looking' at the diagram with this sort of strategy – identifying specific features or substructures that contain *as many elements* as the term number itself, or that occur *as many times* as the term number. This is a useful distinction to be aware of as it has the potential to evoke further, and different, visual apprehensions. By way of example, consider the two pictorial terms illustrated in Figure 5.5. Two

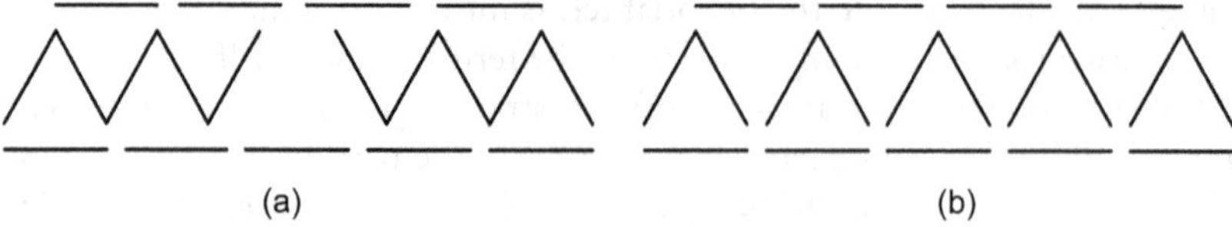

Figure 5.4 Philip's different apprehensions for the expression $T_n = n + 2n + (n-1)$.

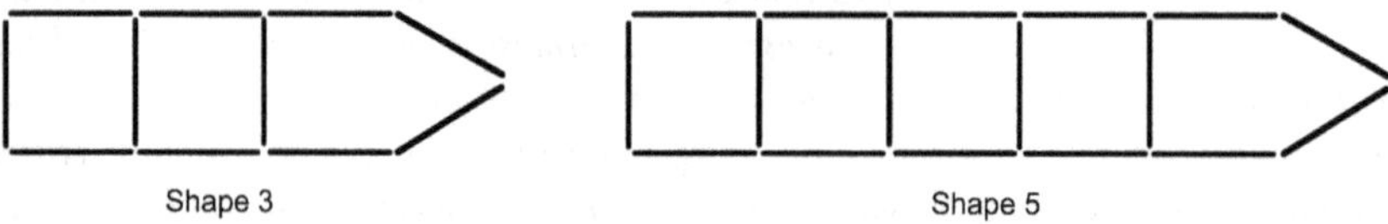

Figure 5.5 Two non-consecutive terms of a pictorial sequence.

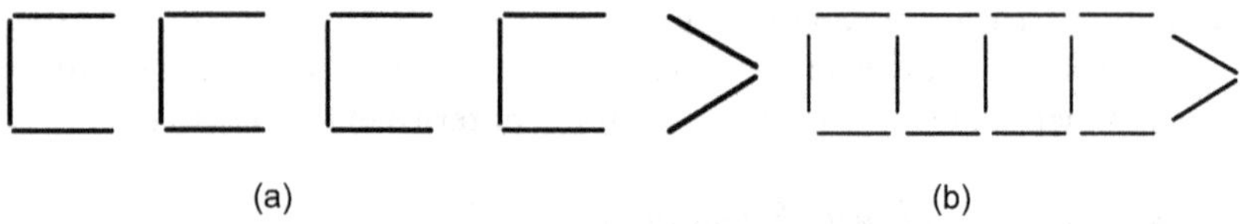

Figure 5.6 Two different visual apprehensions for the expression $T_n = 3n + 2$.

different visual apprehensions, both of which have the same general expression $T_n = 3n + 2$, are shown in Figure 5.6, using Term 4 as a generic example.

In Figure 5.6a the $3n$ part of the expression represents n multiples of three matchsticks (the series of C-shapes), while in Figure 5.6b the $3n$ represents three multiples of n matchsticks (a horizontal row at the top, the vertical row in the middle and a horizontal row at the bottom). Figure 5.6a can be thought of as a sequential apprehension, or 'local generalisation', in that it foregrounds the three-match additive unit that needs to be added to (or inserted into) a term to form the next term in the sequence. By contrast, Figure 5.6b represents a more 'global generalisation' in that it foregrounds substructures that are not brought forth by a consideration of the step-by-step process of constructing one term from the previous one in the sequence. This is a useful distinction to be aware of, as it can be used as a pedagogical strategy for shifting pupils' focus away from the additive unit and towards a more holistic visualisation of potential generalities.

Strategy 3: Focusing on Even-Numbered or Odd-Numbered Terms

Another interesting insight that arose from the micro-analysis relates to the inclusion of either even-numbered or odd-numbered terms in the visual stimulus

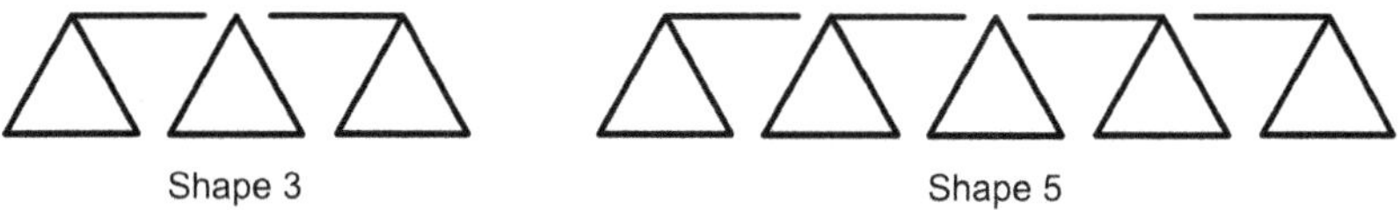

Figure 5.7 Philip's visual apprehension of his general expression $T_n = 3 + 4(n-1)$.

itself. An instance of the juxtaposition of two odd-numbered terms bringing forth specific apprehensions arose with Philip's engagement with the visual stimulus illustrated at the top of Table 5.1. Figure 5.7 shows one of Philip's visual apprehensions in which he used the strategy of "*picking the centre triangle as the sort of cornerstone*".

A crucial trigger for this particular apprehension is the commonality in the symmetry of the two juxtaposed odd-numbered terms. As a prompt I asked Philip if he thought his visual reasoning would still hold for even-numbered terms. He was poised to draw Term 2 when he realised that there was no central triangle as before. He managed to resolve this problem by drawing *inverted* images for Term 2 and Term 4, each of the inverted images now having a central triangle to act as a 'cornerstone' as before (Figure 5.8).

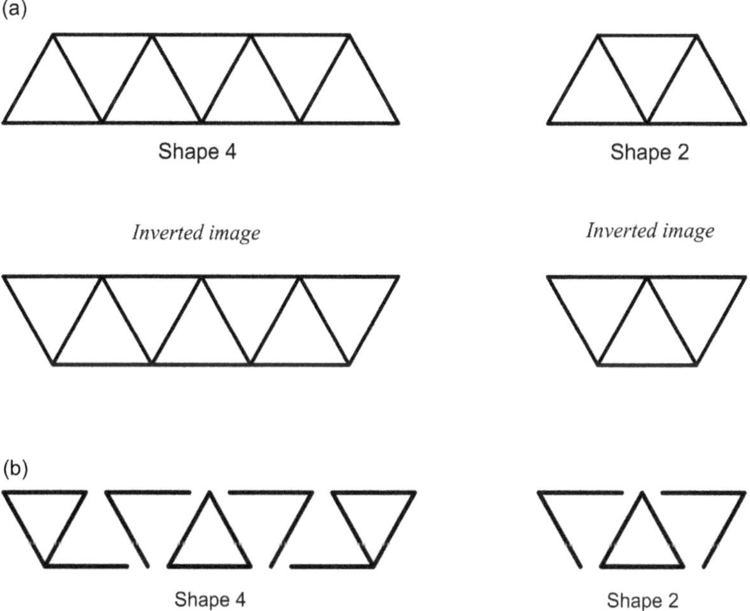

Figure 5.8 Philip's visual resolution for even-numbered terms.

Operative apprehension was thus invoked by a reconfiguration of the whole-part relation of the pictorial terms through a variation in their orientation. His visualisation for *odd*-numbered terms, based on his general expression $T_n = 3 + 4(n - 1)$, is of a central triangle symmetrically surrounded by four-match units as illustrated in Figure 5.7. The complication for *even*-numbered terms is that there are now an odd number of four-match units to symmetrically position around the central triangle. Philip's solution to this problem was to split one of the four-match units into two sections, thereby retaining the overall symmetry. In his own words, "*the first group of four in the even ones just fills up the gaps*" (Figure 5.8).

The manner in which Philip managed to extend his visualisation of odd-numbered terms to incorporate even-numbered terms is remarkable. Not only did it require a complete reconfiguration of the whole-part relation, but this was done in a way that retained generality and conformed to the algebraic expression he had arrived at for odd-numbered terms. For certain pictorial patterns, focusing on only the odd-numbered or even-numbered terms may be useful not only with respect to the generalisation process itself, but also in terms of its potential educational value.

Strategy 4: An Awareness of Embodied Processes of Visualisation

A broad spectrum of embodied processes was exhibited by the research participants. This richly textured tapestry of activity included the use of words, linguistic devices, gestures, rhythm, graphics and physical artefacts. This activity in many instances provided evidence of unconscious visual apprehensions or visualisations that had not yet led to a stable form of awareness. While a conscious search for structure is a useful generalisation strategy, so too is the process of unstructured exploration and interaction with the pictorial context. It is the tactile and whole-body engagement of such activity that has the potential to lead to the serendipitous awareness of structural regularity and lead to crucial pivots for evoking 'new ways of seeing'. Drawing pupils' attention to their own unconscious activity is thus a useful classroom strategy and resonates with the heuristic strategy *Watch What You Do* (Mason et al., 2005), the underlying principle of which is that paying attention to one's engagement with *particular* cases may lead to an awareness of a *general* structure that was not initially apparent.

When presented with her pictorial pattern (the same as Grant's, shown at the top of Table 5.1) for the first time, Kelly counted the matchsticks in Term 3 as illustrated in Figure 5.9a. She then double-checked her answer by re-counting them, but using a different counting procedure as illustrated in Figure 5.9b. In both cases she counted aloud while pointing to each matchstick in turn with her pencil. The first of these counting procedures could be called *economical* in the sense that it minimises the overall distance traversed while pointing to each matchstick in turn. By contrast, the overall distance traversed by the second counting procedure is significantly longer, and as such is *uneconomical*. Such a systematic yet uneconomical counting procedure suggests an underlying visual apprehension, whether conscious or unconscious, on the part of the counter.

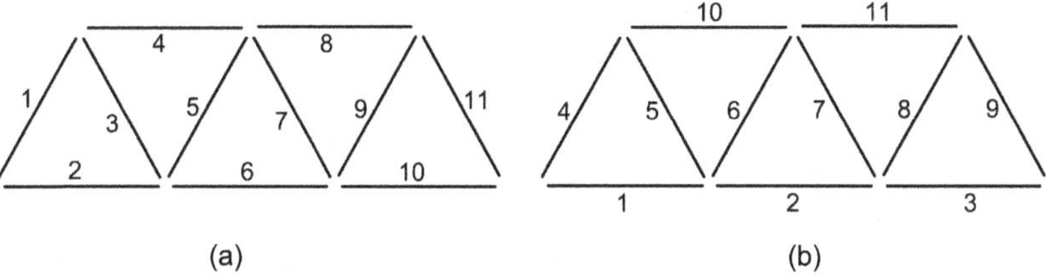

Figure 5.9 Kelly's different counting procedures.

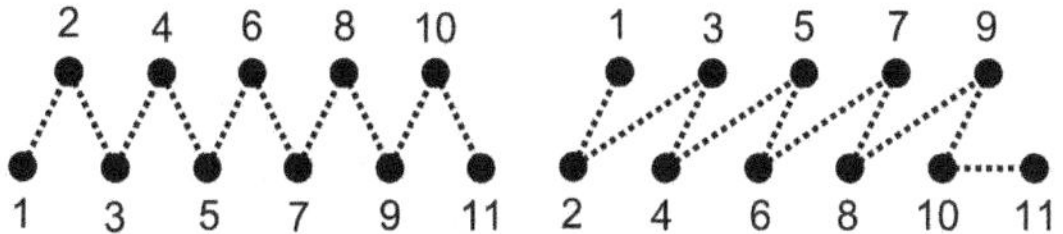

Figure 5.10 Anthea's different counting procedures.

Making pupils aware of how they engage with specific terms has the potential to lead to a more general structural awareness – in this instance to two rows of horizontal matchsticks and a central zigzag of oblique matchsticks.

A further example is illustrated in Figure 5.10 which shows two different counting methods employed by Anthea during her engagement with a pictorial term. The second counting procedure is not only uneconomical, it also gives rise to a sense of *rhythm* as a result of the top-to-bottom path length being shorter than the bottom-to-top path length, and thus being traversed in a slightly shorter time: 1, 2…3, 4…5, 6…7, 8…9, 10…11. An important point here is that the rhythm is an artefact borne out of the counting process itself, an artefact which may lead to the development of structural awareness – in this case the pairs of 'forward slanting' dots and the single unpaired dot at the end. There is a subtle interplay here between the counting process *itself* bringing forth a new apprehension while at the same time playing a crucial role in stabilising the particular visualisation and in developing an awareness of its generality.

Strategy 5: Incremental Steps towards an Expression of Generality

Once a stable visual apprehension has been brought forth and the generality of the visualisation established, the next step is expressing the generality in symbolic form. Verbal expressions such as 'one less than the shape number' and 'two more than the shape number' require a semiotic contraction into the form $(n-1)$ and $(n+2)$, respectively. Such semiotic contractions require the exclusion of linguistic terms that convey important spatial and positional terms, and some pupils may find it useful to approach this semiotic contraction in a stepwise manner. The micro-analysis revealed that some pupils do this unconsciously; for example,

Anthea, who rather than saying "*I subtracted one less than the shape number*", used the more succinct contraction "*I minused the number minus one*". Even her earlier reference to "*the number of the shape*" gets contracted simply to "*the number*". The semantic content of the expression "*the number minus one*" is much more closely aligned with the algebraic syntax ($n - 1$) when compared with the wordier expression "*one less than the shape number*", and this verbal contraction may well be a useful transitional stage en route to desubjectified algebraic symbolism.

In terms of easing the transition to algebraic symbolism, and to avoid unnecessary obfuscation, teachers should take care not to make use of diagrams in which the term number also represents the number of elements in substructures that are likely to be brought forth by pupils. For instance, if there is a likelihood of pupils focusing on squares or other four-unit structures in a particular sequence, then avoid Term 4. Similarly avoid Term 3 if it contains triangular substructures or other three-unit features that could act as visual triggers. This should help in avoiding confusion arising from situations where the same numerical value represents different conceptual aspects of the given pictorial term.

For some research participants, although they were able to visualise the pictorial terms in different ways, the challenge was the process of becoming aware of *how* the visualisation was regular, and how this regularity could be expressed algebraically. Participants exhibited a number of interesting strategies while grappling with this issue. Grant, for example, tabulated a summary of structural features along with the total number of occurrences of each feature for a series of consecutive terms. For Grant this was a powerful way of making sense of the regularity of each of his subdivisions of the pictorial terms. For other pupils a useful strategy was to draw or construct much larger terms than those initially given. Lance, for example, made extensive use of Term 8, while Liza made use of Term 9 on a number of occasions. By engaging with larger terms there is also less likelihood of the same numerical value representing different conceptual aspects of the given pictorial term.

Concluding Comments

Language and action are not simply outward manifestations of internal workings, but rather visible aspects of embodied understanding (Davis, 1995). Semiotic means such as gestures, rhythm and speech are not simply epiphenomena, but they play a fundamental role in the formation of knowledge. Micro-analyses of pupils' whole-body engagement with visual patterning tasks revealed generalisation approaches that evolved out of a purposeful and conscious search for structure, but they also revealed approaches and visual apprehensions that emerged and developed serendipitously from unstructured exploration and interaction with the pictorial context. Quite apart from this pedagogic or epistemological importance, there is a strong moral or ethical dimension to embracing a multisemiotic notion of embodied understanding. This ethical dimension relates to an awareness of, and sensitivity to, the idea that different pupils have different learning styles, different ways of engaging with or making sense of mathematical situations and different ways of *seeing* the world. As Mason (2003) suggests, learning can be seen not only as shifts in the structure of attention, but more broadly as "extending sensitivity to notice" (p. 24). A multi-representational view of pattern generalisation and an appreciation for the cognitive significance of embodied processes thus resonate with the important notion of epistemological access –in terms of how pupils might reflect on their own embodied processes of mathematical understanding, but also in relation to how teachers might gain a deeper and more nuanced insight into their pupils' mathematical sense-making processes.

References

Davis, B. (1995). Why teach mathematics? Mathematics education and enactivist theory. *For the Learning of Mathematics*, *15*(2), 2–9. http://www.jstor.org/stable/40248172

Duval, R. (1995). Geometrical pictures: Kinds of representation and specific processings. In R. Sutherland & J. Mason (Eds.), *Exploiting mental imagery with computers in mathematics education* (pp. 142–157). Berlin: Springer.

Fischbein, E. (1993). The theory of figural concepts. *Educational Studies in Mathematics*, *24*(2), 139–162. https://doi.org/10.1007/BF01273689

Mason, J. (2003). On the structure of attention in the learning of mathematics. *The Australian Mathematics Teacher*, *59*(4), 17–25.

Mason, J., Graham, A., & Johnston-Wilder, S. (2005). *Developing thinking in algebra*. London: Paul Chapman Publishing.

Radford, L. (2006). Algebraic thinking and the generalization of patterns: A semiotic perspective. In S. Alatorre, J. L. Cortina, M. Sáiz, & A. Méndez (Eds.), *Proceedings of the 28th Annual Meeting of the North American Chapter of the International Group for the Psychology of Mathematics Education, Volume 1* (pp. 2–21). Mérida, Mexico: Universidad Pedagógica Nacional.

Radford, L. (2008). Iconicity and contraction: a semiotic investigation of forms of algebraic generalizations of patterns in different contexts. *ZDM Mathematics Education*, *40*(2), 83–96. https://doi.org.10.1007/s11858-007-0061-0

Rowlands, M. (2006). *Body language: Representation in action*. Cambridge, MA: Bradford.

6 Visualisation and Reasoning in Solving Word Problems

Beata Dongwi

Introduction

Numerous studies have been conducted to analyse various aspects of mathematical reasoning, diagrammatic reasoning (Sochański, 2018), geometric reasoning (Mamolo et al., 2015) and visual reasoning (Prusak et al., 2012) among others. Other studies have also been conducted to analyse visualisation as a problem-solving and a reasoning tool in mathematics (Arcavi, 2003; Presmeg, 1986; Rivera et al., 2014). However, research that focuses specifically on the exact relations between the two processes has been elusive, hence this chapter. Much research has focused on understanding reasoning and visualisation processes in parallel (Brodie, 2010; Lithner, 2008a; Natsheh & Karsenty, 2014), but not on the two processes' intertwined inseparability. The research described in this chapter analyses the relations between reasoning and visualisation processes when selected Grade 11 students, working in small groups, solved given mathematical problems.

The study was conducted before the Namibian Broad Curriculum review of 2018. Prior to the 2010 curriculum review, Namibian Grade 12 learners could opt to either take mathematics or not. However, when the transition from Cambridge to a Namibian curriculum was made in the 2010 curriculum review, the concept of education for all was adopted (Namibia. Ministry of Education [MoE], 2010) and all learners were expected to do mathematics from pre-grade to Grade 12. High school mathematics was taught at three levels: core level, extended level and higher level. The core level consisted of less abstract mathematics, whereas the higher level consisted of complex mathematical topics such as differentiation, integration, vectors in 3D, trigonometry, etc. The extended level ranged between core and higher levels. Learners could then opt to enrol for one of these three levels, depending on their mathematical abilities.

Thus, the mathematics curriculum prior to 2018 made it easier for learners with different mathematical abilities to choose the level of mathematics which they deemed doable. However, post the 2018 curriculum, learners are no longer able to choose between these different levels of mathematics. There is now only one compulsory strand of mathematics for all up to Grade 11 and only mathematically inclined and able learners proceed to do mathematics at an Advanced

DOI: 10.4324/9781003172420-8

Subsidiary Level (AS Level) in Grade 12. This has created many challenges and, of course, opportunities whose discussion falls beyond the scope of this chapter.

The word problems selected for this particular research study, however, consisted of problems that could be solved by learners taking mathematics at any level: core, extended and higher level. The content of the tasks was embedded within the students' everyday contexts and thus did not discriminate according to their mathematical level or ability.

Before discussing the empirical findings, the conceptual and theoretical frameworks used in the research study are described.

Visualisation and Reasoning Processes

When working with the participant learners, both visualisation and reasoning processes (RPs) were observed when they attempted to solve a set of given mathematical problems. These processes were both mental and physical, depending on the observers' interpretations of the observed situation and actions. According to Simmt and Kieren (2015), the observer is one "who arises in the act of observing and whose knowing is explained through the mechanism she [or he] describes" (p. 307). I argue in this chapter that there is a close connection between visualising and reasoning and that the two cannot really occur independently of each other. The research challenge was to find evidence as to how visualisation processes were manifested in RPs and vice versa. How do these two processes relate to each other in the context of word problem-solving?

Mathematical Reasoning

Learning to communicate and to reason is crucial for learners in all mathematics classrooms. The Namibian mathematics curriculum foregrounds thinking and reasoning as one of the important skills of learning and doing mathematics. The curriculum recommends that mathematics should provide learners with opportunities to develop essential skills such as communication, problem-solving, and critical and creative thinking, amongst others (Namibia. Ministry of Education, Arts and Culture [MoEAC], 2018). The curriculum further emphasises that learners should "use mathematics as a means of communication with emphasis on the use of clear expression" (Namibia. MEAC, 2018, p. 5), and to "devise mathematical arguments and use and present them precisely and logically" (ibid.). However, despite the accuracy of their responses to mathematical problems, many learners still struggle to reason mathematically (Dongwi, 2018).

Brodie (2010) underscores communication as an integral role in the mathematical reasoning of both individual students and students working in small groups. This kind of reasoning also depends on some key players such as the nature of the task, the solver's worldview and the social context. Boesen et al. (2010) concur and observe that the reasoning used when trying to solve a task depends not only on the nature of the task, but also on the relation between the task, the solver and the social context. The kind of reasoning that a task elicits is thus important for "the students' learning since this is the reasoning that feedback can address" (Boesen et al., 2010, p. 90).

Moreover, reasoning is defined as the line of thought adopted to generate assertions and conclusions when solving mathematical tasks (Lithner, 2008b). According to Tripathi (2008), one of the ways to promote mathematical reasoning among learners is to encourage them to use visual representations in a problem-solving context. This is essential because it enables them to interpret, communicate and discuss their ideas with others.

Reasoning processes

Patterns of reasoning in the learners' responses to geometry word problems in this chapter were classified in terms of the following RPs: explanation, justification, argumentation and generalisation.

Explanation (RPE) refers to the clarification aspects of someone's mathematical thinking that he/she thinks might not be readily apparent to others. Webb (1991, p. 368) notes that content-related explanations consist of descriptions of how to solve a problem or part of a problem that includes some elaboration of the solution process.

Justification (RPJ) is a means by which "students enhance their understanding of mathematics and their proficiency at doing mathematics; it is a means to learn and do mathematics" (Staples et al., 2012, p. 447). The purpose of a justification in the context of word problems is to provide a convincing argument, such as justifying why carrying out a series of representations is a valid method for determining the answer to a given word problem.

Argumentation (RPA) is defined as the "substantiation, the part of reasoning that aims at convincing oneself, or someone else, that the reasoning is appropriate" (Lithner, 2000, p. 166). It is a sequence of statements, sentences or formulae, such that each is either a premise or the consequence of previous lines of argument, and the last of which is the conclusion (Dove, 2009). It involves refutations and/or acceptances of claims and justifications.

Generalisation (RPG) means to introduce new ideal objects, to overcome objective constraints (Otte et al., 2015, p. 144), to identify the operators and the sequence of operations that are common among specific cases and to extend them to the general case (Swafford & Langrall, 2000, p. 91). A generalisation of a problem-solving situation may be presented verbally or symbolically.

Visualisation

Visualisation in this chapter is defined as the process of transforming a mathematical idea into a diagram or a picture (Hershkowitz et al., 2001; Makina,

2010). Van Garderen et al. (2014) observed that diagrams could also be "powerful ways to facilitate communication about critical ideas in mathematics as well as provide a platform for sharing problem solving strategies with others" (p. 136). Thus, "helping students to become aware of the importance of making drawings in mathematics problem solving" (Csíkos et al., 2012, p. 62) will enable them "to engage with concepts and meanings which can be easily bypassed by the symbolic solution of the problem" (Arcavi, 2003, p. 222). Natsheh and Karsenty (2014) concur that visualisation is the process of communicating information, critical thinking, developing ideas and advanced understanding in solving mathematical problems through using images, pictures and diagrams. Therefore,

> [b]y saying farewell to the drilling practice in the world of word problems, we may offer new resources for efficient classroom methods that help students become aware of their mental processes and of the importance of using appropriate visualisation methods in solving word problems.
>
> (Csíkos et al., 2012, p. 62)

Visual Imagery

According to Presmeg (2014), it is imperative that mathematics teachers encourage their students to engage in visual thinking and reasoning. Visual imagery is used in this chapter to define the act of visualising during problem-solving. Hegarty and Kozhevnikov (1999) define visual imagery as the ability to form mental representations of the appearance of objects and to manipulate these representations in the mind. One can generally say that we think visually of a mathematical concept if we tend to visualise a diagram associated with that particular concept (Sochański, 2018). Therefore, visualising objects and graphically representing numerical information are important mathematical tools that help learners to solve problems and to understand complex mathematical concepts. The study described in this chapter adapted Presmeg's (1986) categories of visual imageries (VIs) to observe and analyse the extent to which the research participants opted to use visual imagery to solve selected word problems. These categories of visual imageries are defined as follows:

Concrete Pictorial Imagery (CPI) – this refers to the concrete image(s) of an actual situation formulated in a person's mind – i.e. a picture described verbally, drawn on paper or using a technological device.

Pattern Imagery (PI) – this refers to the type of imagery in which concrete details are disregarded and pure relationships are depicted in a visual-spatial scheme. The essential feature of PI is that it is pattern-like and stripped of concrete detail (Presmeg, 1986).

Memory Imagery (MI) – this refers to the type of imagery that enables visualisation of an image that one has seen somewhere before. MI includes a history of recurrent occurrences.

Kinaesthetic Imagery (KI) – this is the kind of imagery that involves muscular activity. A kinaesthetic visualiser wants to feel and touch.

Dynamic Imagery (DI) – this imagery involves processes of transforming shapes, i.e. redrawing given or initially drawing one's own figures with the aim of solving the problem.

Enactivism

An enactivist perspective of human cognition underpins the study described in this chapter. Enactivism, as defined by Begg (2013), is "a way of understanding how all living organisms including human beings, organise themselves, and interact with their environments" (p. 18). According to Maturana and Poerksen (2004), an enactivist perspective questions how the elements of a system work together to form that system. This chapter reports on a study that was conducted to investigate how selected Grade 11 learners interacted with themselves, each other and the environment, to solve word problems and reach a collective solution. Brodie (2010) underscores communication as an integral role in the mathematical reasoning of both individual learners and learners working in small groups – a notion embedded in the very nature of the enactivist perspective. The study thus observed how learners communicated verbally and nonverbally, i.e. through gestures and other visual forms, as they reasoned about their problem-solving strategies in their small groups – availing an opportunity to analyse the co-emergent relationship between visualisation and RPs.

Co-emergence is an enactivist concept interpreted by Li et al. (2010, p. 407) as a change of both: a living system and its surrounding environment, depending on the interaction between the system and its environment. When the system and the environment interact, they become structurally coupled. This means that the mutual interaction of the organism and the environment causes changes and transformations in both (Khan et al., 2015). Begg (2013) claims that from an enactivist perspective, humans and the world are inseparable: they co-emerge. As a result, cognition (learning) cannot be separated from being (living), and knowledge is the domain of possibilities that emerges as we respond to and cause changes within our world. Thus, one cannot separate knowing from doing and from the body. Brown (2015) stresses that "we are co-emergent and where there is a coordination of actions, like in a classroom, or a collaborative group in a research project, a culture of practices emerges that is good-enough (effective action) to get done what needs to be done" (p. 188). Li et al. (2010) caution that while co-emergence suggests that the system and the environment interact, it does not guarantee greater or lesser adaptation on the part of either to each other. Students bring forth a world; they emerge with it, but it is their structures that bring them forth (Proulx, 2008, p. 22).

According to Khan et al. (2015), enactivism is attentive to the coupling of organisms and their environments as it attends "explicitly and deliberately to action, feedback and discernment" (p. 272). Furthermore, "enactivism emphasises on knowing rather than knowledge" (Begg, 2013, p. 84), and it is concerned with "*learning in action*" (Khan et al., 2015, p. 272). Hence, enactivism acted as a linking mediator to bring visualisation and RPs together as discussed in the next section.

Research Methodology

The study reported in this chapter took the form of a case study research that was conducted within an interpretive paradigm, which endeavours to understand the subjective world of human experience (Cohen et al., 2011). Anchored by the concerns of understanding human agency, behaviour, attitudes, beliefs and perceptions (Bertram & Christiansen, 2014), the interpretive paradigm fits an enactivist study.

In the context of groups of selected learners interacting with each other when solving given word problems, the overarching research question that framed this chapter is, how do reasoning and visualisation relate to each other when selected Grade 11 learners solve mathematical word problems in small groups?

Participants and Data Collection

A total of eight learners participated in the second phase of a bigger study (Dongwi, 2018; Dongwi & Schäfer, 2019), which is the focus of this chapter. The participants were purposefully selected for their preferred visual methods when solving mathematical problems, when the algebraic methods they favoured were compared. This was determined from the results of the first phase of the study. Part of the selection process also included a short questionnaire given to selected teachers at the school who knew the participants quite well. The purpose of this questionnaire was to determine which participants could work well in pairs or in small groups. The participants in this chapter came from two small focus groups: Group one consisted of Ethray and Ellena, and Group two of Denz and Jordan (not their actual names).

Data were collected using focus-group, task-based interviews where each group solved given geometry word problems in the presence of the researcher, whose role was that of a participant observer. Each group was provided with a single worksheet consisting of five geometry word problems that the participants were tasked to solve and to then reach a collective solution. They solved the problems both on the worksheets, and through articulations that could not be scribbled on paper. The participants also engaged bodily actions such as gestures spontaneously to explain and support their answers, justify their claims and accept or refute others' arguments. These actions were also used to support and reinforce their generalisations. The interviews were video recorded to ensure that both verbal and nonverbal articulations are transcribed and analysed.

Data Analysis

To answer the research question, data from each of the focus-group interviews were analysed, using both visualisation and RP frameworks (see Table 6.1).

Interview transcripts were uploaded into *NVivo* (a qualitative coding and analysis software package developed by 'QSR International') for data coding, to enable the analysis of the co-emergence of visualisation and RPs. Instances of each visual imagery and RP in each transcript were coded using observable indicators in their respective analytical frameworks (Table 6.2 shows the theoretical frameworks). The *NVivo* software enabled coding of a single segment to be more than one type of VI or RP. This is because a learner could use more than one type of VI at the same as engaging more than one RP, as was often observed during data analysis (shown in Table 6.1). The quotes illustrated in Table 6.1 are thus selected from the vignettes analysed in this chapter, and, to a lesser extent, provide the reader with an idea of the data coding in the transcripts.

Thereafter, a fine-grained analysis in the form of a series of selected vignettes helped with a more meticulous and detailed discussion of the data, for the co-emergence of visualisation and reasoning during the second phase of the study. In this chapter, four vignettes are analysed and discussed. The analysis centred on the actions, gestures and arguments that individual research participants 'uttered'

Table 6.1 Analytical framework 2 – reasoning processes template

Reasoning Processes (RP)	*Code*	*Observable indicators/researcher's inferences* *A learner:*
Explanation	**RPE**	**RPE1**: makes sense of the problem and establishes a claim, e.g. explains what the problem entails in simple terms and suggests known concepts/procedures **RPE2**: explicates his/her own thinking processes (to produce meaning – includes reasoning without words) **RPE3**: suggests and defines problem-solving strategies
Justification	**RPJ**	**RPJ1**: provides proofs to validate claims and arguments **RPJ2**: provides acceptable reason for action (asks for clarification from others) **RPJ3**: promotes understanding among those engaged in justification, e.g. does something to answer another person's concerns and lessen their worries
Argumentation	**RPA**	**RPA1**: provides support for explanations and justifications (this includes insisting on accuracy of their own and others' claims) **RPA2**: convinces/persuades others via verbal/visual activity of the truth of their claims and appropriateness of their reasoning (or is convinced and persuaded by others – i.e. when they accept the truth of each other's claims and explanations) **RPA3**: accepts/refutes truth of others' claims that they may agree/disagree with
Generalisation	**RPG**	**RPG1**: elaborates the problem further to try to learn more from the result by relating the problem to similar situations **RPG2**: uses visualisation to demonstrate how the problem can be solved in a different way

Table 6.2 Analytical framework 1 – visual imagery

Category of visual imagery	*Code*	*Observable indicators/researcher's inferences A learner:*
Concrete pictorial imagery	**CPI**	**CPI1**: formulates a picture in the mind (PIM) (while reading/ rereading a word problem); draws/sketches to represent a mental image or a concrete situation **CPI2**: concentrates silently (after a question is posed) – the thinking process involves imagination and mind pictures **CPI3**: clarifies the structure of the problem; gives explanations/ suggestions based on imagination and the formulated PIM
Pattern imagery	**PI**	**PI1**: formulates/uses patterns with the purpose of depicting/ communicating information. For example, patterns of the theorem of Pythagoras (i.e. $c^2 = a^2 + b^2$) **PI2**: engages patterns of data and arguments **PI3**: uses visualisation to discover generalisations and to derive nonobvious concepts/formulae from such generalisations
Memory imagery	**MI**	**MI1**: formulates a mental image of a book/ board and depicts how a formula/concept was written (visualises something previously learned) **MI2**: sees a specific formula/method in mind that is needed to solve the problem (he/she may give description of the problem-solving strategy) **MI3**: recalls from memory; uses previous knowledge; an act of remembrance
Kinaesthetic imagery	**KI**	**KI1**: patterns of movement and body engagement as part of problem-solving **KI2**: walks/traces a path with fingers/hand/pencil to illustrate an image of something **KI3**: mimics/imitates/traces shapes without placing the pencil on paper
Dynamic imagery	**DI**	**DI1**: redraws given or own drawn diagrams with a purpose of extracting simple figures from complex figures, or to divide figures with lines to form other figures **DI2**: visualises a series of several images connected in one smooth motion, in mind and/or paper **DI3**: transforms/changes the orientation of picture/shapes/ concrete objects

when they solved geometry word problems in their small collaborative groups. Although the fine-grained analysis focused on individual participants' mathematical reasoning, thinking and visualisation, consideration was also given to the social setting in which they belonged, i.e. their small groups; hence, other participants' influence on their reasoning was carefully considered for each vignette.

These four vignettes were selected from two small groups of participants when they solved two selected tasks. The two tasks were selected because of their embedded inherently visual nature. Both word problems were presented to the participants without accompanying diagrams, which enabled and encouraged the participants to use their own visual representations from the outset. The participants working in these two groups were selected for this chapter because they worked well together (as illustrated in the vignettes). Their structures coupled throughout the problem-solving process – for example, they often completed each other's sentences, drawings and other visual aspects. Therefore, their inclusion in this chapter allowed for an in-depth analysis of the co-emergence of visualisation and RPs in word problem-solving. Although visualisation and RPs were observed in all the scenes and tasks during task-based interviews, the selected vignettes were quite particular in terms of how the participants solved the problems together. These four vignettes were thus considered for their striking visual representations and verbal and nonverbal reasoning, by both groups.

Two Tasks, Two Groups, Four Vignettes

Task 1: Mr. Onesmus has a plan for his sick goat. He tied the goat to a tree with a 7 m rope in such a way that the goat is able to move freely around the tree for it to graze. If the goat moves a complete revolution with the maximum length of the rope, what is the total possible area that the goat would graze?

Consider the following two vignettes, analysing the reasoning and visualisation processes of the two sets of students when they solved Task 1 in their small group.

Vignette 1: Denz and Jordan

In this vignette, Denz and Jordan engaged the RP of argumentation (RPA) more than the other RPs. Their mathematical arguments were coupled with different aspects of VIs, which intertwined of course with more reasoning processes. It should be noted here that the coding inserted inside parentheses to represent either VI or RP is simply an example of how data were coded in the bigger study. In these vignettes, many of the VIs are written (not coded) in square brackets, often after the participants' direct quotes, while many of these quotes were not coded for reasoning processes.

Jordan started reading the task and Denz immediately started sketching as he listened to Jordan (CPI). Denz sketched a tree and a rope (Figure 6.1a). When Jordan realised that Denz was sketching, he slowed down his reading pace and even paused at times to allow Denz to complete his sketch. Denz then gestured a circular movement with his hand to show the grazing area of the goat (KI, DI). He said, "*So, here is the tree, you can see the tree is in the middle*" [points to the sketch as he speaks]. Jordan continued reading from where he paused and Denz sketched another diagram, a circle with a centre and a radius of 7 m (Figure 6.1b) (DI). Both boys then read the last part of the question slowly, with their pencils pointing at the diagram. "*Okay, so, here is a tree, you can see the tree is in the middle*", said Denz as he marked off the centre of the circle, which represented the tree in the first sketch.

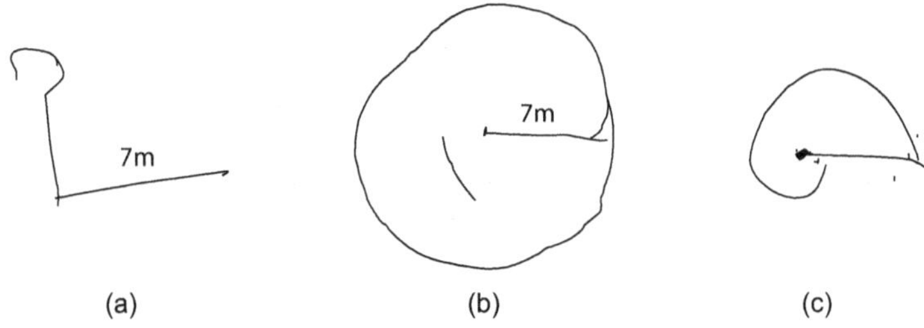

Figure 6.1 Denz' and Jordan's visual representation of Task 1.

In order for the goat to graze freely, he should be able to move the maximum distance and area around the tree, which is represented by the locus of points about a fixed point, or simply a circle of radius 7 m. Jordan, however, argued that as the goat moves around the tree, the rope gets shorter. Here is a short excerpt from the transcript. Visualisation processes are in square brackets (RPA).

JORDAN: If it goes around, if it does a revolution, the rope gets smaller. [Sketches a diagram of what he means with the rope getting smaller].
DENZ: Will it get smaller?

...

DENZ: No, no, no... Remember this is standing up straight and...this... [mimics a standing up tree with an upward movement using a pencil].
JORDAN: Yeah, the tree is going to be up, you know...
DENZ: 7 so it can go 7 m around all the way...or...!
JORDAN: Yeah, it gets shorter when it *mos ties around to*...in such a way that the goat is able...

By the rope getting shorter, Jordan meant that the rope would wind itself around the tree trunk if the goat moved in a circular motion (Figure 6.1c), but Denz convinced him that if the rope was loose enough it would move freely around the tree trunk (RPA).

The learners' reaction to this task was both visual and motivated by the reasoning behind each visual aspect, be it physical or mental imagery. Both verbal and nonverbal reasoning were taken into consideration as they shaped the problem-solving strategies and also contributed to the participants' solution.

Vignette 2: Ellena and Ethray

This vignette presents Ethray and Ellena's explanations (RPE) when they solved Task 1 in their small group. This vignette was selected because both

Ellena and Ethray used the reasoning process of explanation more extensively to solve this task during the task-based interview, in comparison to the reasoning processes of justification (RPJ) and argumentation (RPA).

As soon as he finished reading the task, Ethray swiftly unpacked what the problem entailed (RPE), made suggestions (RPA) and sketched the diagram (RPJ, CPI) all at once. He said, "*Okay, so, there's a tree and there's 7 m rope*", and sketched a tree and a 7 m line from its trunk as illustrated in Figure 6.2a (CPI). In Figure 6.2b, Ellena, seeing a different picture in her mind (CPI), a rather more geometric picture in comparison to the one that Ethray had drawn, asked him: "*Why don't you just do this for the tree* [marks a point – the centre], *and this for the rope?* [draws a line from the point – the radius of the circle] (CPI, RPE) *Then you can just...*" [Mimics drawing a circle with the tree as the centre and the rope as the radius].

In Figure 6.2, Ethray illustrated what was observed as a real-life image of an actual situation in his mind, (a) while Ellena enacted a mathematical transformation of the image of the same situation (b). Ethray accepted Ellena's suggestion of a more geometric drawing when he verbalised as follows (RPA):

> Seven meters okay. So, the goat can graze freely around it [walks a circular path with a pencil to show the boundary where the goat will graze] if he wants to (...) if the goat moves a complete revolution, what is the maximum

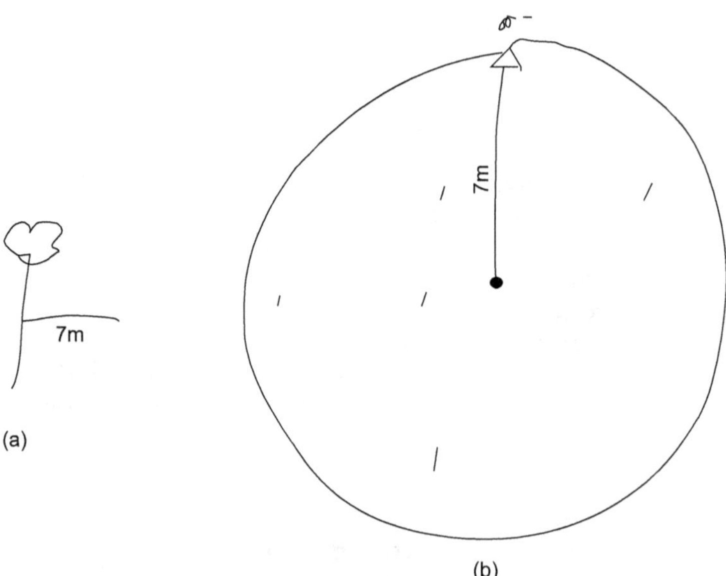

Figure 6.2 Ethray's (a) and Ellena's (b) visual representation of Task 1.

length of the rope? What is the total possible area the goat would graze? We have got to find the area of the circle.

Ellena agreed that they needed to find the area of the circle and thus started calculating this area while the discussion was still under way. She spoke and wrote at the same time: "$A = \pi r^2 = (7)$" (MI). While Ellena busied herself with elaborating on the formula, Ethray worked on the calculator and provided her with the solution when she needed to write it down. It was observed that the two participants effectively 'read each other's minds' and completed each other's sentences as they reasoned their way through solving the problem. This process of completing each other's sentences and sketches in addition to expressing certain inner thoughts together was observed throughout the task-based interview. Damiano (2012) defines this type of dynamic interaction where an autopoietic system and its environment are structurally coupled as co-emergence: in this context, the co-emergence of visualisation and reasoning processes.

Task 2: A pack of 52 cards is dealt out to 10 people seated around a circular table in such a way that the first person gets the 1st card, the fourth person gets the 2nd card, the seventh person gets the 3rd card, the tenth person gets the 4th card, and the third person gets the 5th card and so on. Which person gets the last card?

Once again, two selected vignettes are presented to analyse the relationship between visualisation and reasoning of the two sets of learners when they solved this task in a group setting.

Vignette 3: Ellena and Ethray

Ethray and Ellena's interactions further coupled and co-emerged when they interchangeably visualised and reasoned during the fourth task. Ellena started reading the task but then turned the worksheet towards Ethray to read instead. Ellena preferred to observe while somebody else read. She followed and made the task visual on paper as she listened to Ethray's reading. She remarked, "*So, there's two, two, two spaces*" [jumps two spaces with her pencil as she realised the pattern in the distribution of cards to the people seated around an imaginary table] "*... but there's supposed to be a formula!*" Ellena recalled from memory (MI) that they could find a formula to help them answer sequence related questions. She had recently learned this in her mathematics classroom but was unsure of how to implement it during the focus group task-based interview. Ethray explained to her that there was more than one way of getting the answer to the question. "*There should be a formula I know but, let's just see if we can get the answer then try to get the formula*" (RPE, RPG, KI) [overt gestures and body movements as Ethray speaks]. Ethray continued, "*So, this is 5, 6, 7, 8, 9, 10, 11, 12, 13, 14, 15, 16, 17, 18, 19, 20, 21, 22, 23, 24, 25, 26, 27, 28, 29*" [Ellena observes attentively as Ethray allocates imaginary cards to imaginary people seated around the table]. Seeing that Ethray was involved in the pattern, Ellena alerted him to skip two people when allocating the cards as this was part of the worked-out pattern. Ellena alerted him, "*You can't just start here, you should skip two and so there is a 1, 2, 3, 4…*". Realising this and accepting his mistake Ethray restarted the process (PI & RPA), and together with Ellena, counted all the numbers, allocating cards to ten imaginary people sitting around an imaginary table (CPI, KI, DI). When I asked whether they were going to count up to 52, Ethray responded: "*Yes, ma'am* [and continued to count] *30, 31, 32, 33, 34, 35, 36, 37, 38, 39, 40, 41, 42, 43, 44, 45, 46, 47, 48, 49, 50, 51, 52. The fourth person gets the last card*", he announced.

Although Ellena arrived at the solution that the fourth person gets the last card at around about the same time as Ethray, she insisted that they needed to get the formula (despite the fact that they initially struggled to formulate it). I asked them to explain how they arrived at their solution and how their strategy could help them to answer similar questions in future (RPG). To this, they constructed a table, which Ethray suggested they would populate until a pattern started to form. He suggested:

> Okay, so the first person gets the 1st card [Ellena allocates imaginary cards as Ethray speaks – see Figure 6.3]. Fourth person gets 2nd card. The seventh person gets third card. Tenth person gets a fourth card. Can I just continue? (…) So, the third person gets the fourth card…no, fifth…and then sixth person gets the sixth card. And the ninth person gets the seventh card. Second person gets the tenth card. Then you can see a pattern forming here (…).

Figure 6.3 illustrates how Ellena visually represented Ethray's long description above. This happened at the same time – Ellena visualised while Ethray reasoned.

1	2	3	4	5	6	7	8	9	10
1st			2nd			3rd			4th
		5th			6th			7th	
	8th			9th			10th		
11th			12th			13th			14th
		15th			16th			17th	
	18th			19th			20th		
			22nd						

Figure 6.3 Ellena's and Ethray's visualisation of Task 2.

The participants worked beautifully together here. The two learners' thinking and actions synchronised; hence their visualisation and reasoning processes for this task displayed co-emergence at every instance.

To the question of why the fourth person received the last card, Ethray answered as follows:

> I think the reason the fourth person gets the card is because there are 52 cards. The fourth person will receive the second card, if you add ten the whole time (PI) till you reach twelve...now which person got the second card? You end at four because there's not enough cards (...) Like if you look at a person...if the cards were 72, which person would get it? Would it still be the fourth person? I think it still (...) I think it still will be the fourth person, because of the two.

It is evident from the above excerpt that Ethray formulated some kind of a pattern (PI) in his mind while allocating cards to imaginary people around the table. In an effort to ensure that his reasoning was firm, I persistently asked him to explain his reasoning of the phrase he used "*because of the two*", which he explained referred to "*the two at the end...what the person received*". Ellena also helped to clarify her teammate's reasoning as she commented: "*The second and the twelfth and the ...*" [points at the numbers ending with a two as she follows the pattern] (RPJ). Ethray cut in on Ellena and completed her sentence: "*and the 22nd and the 32nd, and the 42nd, and the 52nd*" [PI].

While Ellena and Ethray opted for a more visually oriented solution, Denz and Jordan diverted from an initial visual representation of the problem to an application of the arithmetic sequence formula. This still falls within the visual imagery framework as memory images of formula or MI (Presmeg, 1986). Although the formula may be derived from visual deliberations of the problem, it remains embedded in the learners' prior knowledge and hence its aspects are mainly derived from memory.

Vignette 4: Denz and Jordan

In this vignette, we observe how Denz and Jordan attempted and solved the pack of 52 cards problem. Denz and Jordan's initial approach to this task resembled that of Ellena and Ethray, the only difference being when they applied the arithmetic progression formula after much deliberation. Figure 6.4a illustrates Denz and Jordan's representation of the word problem (CPI), while Figure 6.4b shows how they used the formula to answer the question (MI).

In Figure 6.4a, the grey box illustrates the supposed ten people seating around the table. The light grey circles indicate the first round of issuing the cards, the

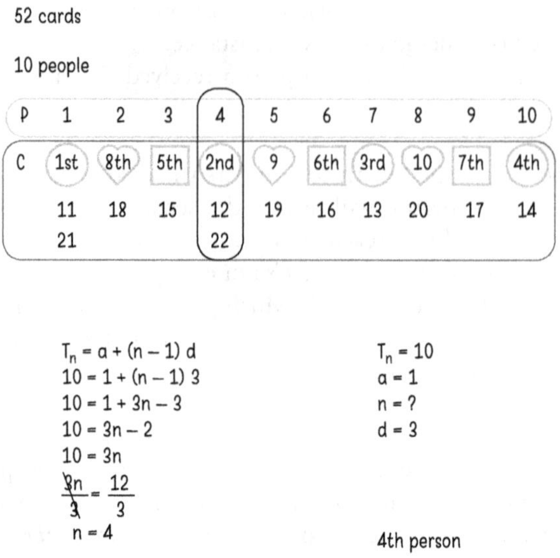

52 cards

10 people

P	1	2	3	4	5	6	7	8	9	10
C	1st	8th	5th	2nd	9	6th	3rd	10	7th	4th
	11	18	15	12	19	16	13	20	17	14
	21			22						

$T_n = a + (n - 1)\,d$
$10 = 1 + (n - 1)\,3$
$10 = 1 + 3n - 3$
$10 = 3n - 2$
$10 = 3n$
$\frac{3n}{3} = \frac{12}{3}$
$n = 4$

$T_n = 10$
$a = 1$
$n = ?$
$d = 3$

4th person

Figure 6.4a Denz and Jordan's visual representation of Task 2.
Figure 6.4b Using the formula to solve the problem.

light grey hearts the second round and the light grey squares the third round. The black oval that cuts through the grey and black boxes illustrates the point of interest – the pattern of who gets the cards of interest.

Coupled with this visual representation are reasoning processes when Denz and Jordan deliberated their way through problem-solving. The transcript excerpt below illustrates the preliminary coupling of their problem-solving strategy.

JORDAN: *Put the person, 1, 2, 3, 4, 5, 6, 7, 8, 9, 10* [visualises the ten people seated around the table by allocating each person with a number from 1 to 10].

DENZ: *So these are people. Okay, so person one gets the first card. The fourth person gets the second card.*

JORDAN: *Second card.* [Allocates cards to people that he visualised in his mind as Denz instructs who gets what card].

DENZ: *The seventh person gets the third card. Tenth person gets the fourth card.*

JORDAN: *Oh, I see, you see a pattern here?*

DENZ: *Mm.*

JORDAN: *They skip two people.*

DENZ: *And the third person gets the fifth card.*

Denz suggested that they could use the formula but neither of them could recall the formula, so they used Figure 6.4a to derive it. I think this was effective thinking from their side and their visual representation of the problem proved useful in helping to engage operational reasoning.

We can easily see how visualisation and reasoning are inseparable during problem-solving here. The boys visualised to help them reason and then reasoned to help them visualise. After figuring out the common difference, the boys used another strategy to ensure that they could generalise their solution (RPG). They supported each other in their thinking and encouraged each other to discover more patterns. This is a strong tenet of enactivism: bringing forth a world of significance with oneself and others in a social and collaborative setting.

The boys worked together and interacted beautifully. The differences between visualisation and reasoning processes were also subtle here. One can hardly separate them as they intricately coupled with each other. Concurring with Maturana and Varela (1998, p. 75), a history of recurrent interaction between these two boys led to the structural congruence between them and the environment in which they were operating. According to Maturana and Varela (1998), this is called structural coupling.

The way that these boys solved the task vindicated my conviction of how visualisation and mathematical reasoning relate and co-emerge with each other. There was a noticeable pattern in how the boys deliberated over the task and arrived at a collective solution. Denz started by reading the question while Jordan studied the given figure to correspond with what Denz read. The boys chorused most of the answers in unison and often completed each other's sentences when they provided reasons for their visual representations. Jordan mostly sketched while Denz read the question, although they swopped these roles with ease. They were both involved in the task at the same time.

Conclusion

The study discussed in this chapter was conducted to analyse the relationship between visualisation and reasoning in the context of word problem-solving. The argument is that these mathematical processes are inseparable, that is, they are intertwined in such a way that they co-emerge. To analyse this co-emergence, selected learners solved given problems in small groups and their enacted visualisation and reasoning processes were observed and analysed. Interview transcripts as well as gestures captured on the videos were coded using the two analytical tools (Table 6.2). Table 6.1 demonstrates an example of the inseparability between these two processes. The findings reveal that there is rich evidence of co-emergence between the two pairs' visual imagery and the mathematical reasoning that both groups displayed in all four vignettes. The analysis of the data presented compelling evidence of the participants constantly completing each other's sentences, suggestions and conjectures during a very collaborative problem-solving process.

Even when their calculations were at times inaccurate and incorrect, the participants still collaboratively argued throughout their problem-solving process, reasoned well and rigorously and worked as a team to rectify that which they deemed inaccurate. The support that the learners rendered each other and the way in which they listened to each other during argumentations and justifications

Table 6.3 The inseparability of visualisation and reasoning processes

	Reasoning processes				
Visual imagery		**Explanation** Clarifying aspects of mathematical thinking	**Justification** Validation of claims to provide insight into the phenomenon	**Argumentation** Acceptability or refutability of a conclusion	**Generalisation** Extension of identified common operators to the general case
	CPI Picture in the mind drawn on paper or described verbally	"*Okay, so, there's a tree and there's 7 m rope*" [sketched a tree and a 7 m line from its trunk] **ETV2** "*why don't you just do this for the tree*" [marks a point – the centre], *and this for the rope* [draws a line from the point – the radius of the circle] **ELV2**	*If it goes around, if it does a revolution, the rope gets smaller.* [Sketches a diagram of what he means with the rope getting smaller] **JV1**	By the rope getting shorter, Jordan meant that the rope would wind itself around the tree trunk if the goat moved in a circular motion (Figure 6.1d), but Denz convinced him that it the rope was loose enough it would move freely around the tree trunk. **DV1/JV1**	Asked them to explain how they arrived at their solution and how their strategy could help them to answer similar questions in future. They constructed a table, which Ethray suggested they would populate until a pattern started to form. **ETV3**
	PI Concrete details are disregarded, and pure relationships are depicted in a visual-spatial scheme	"*So, this is 5, 6, 7, 8, 9, 10, 11, 12, 13, 14, 15, 16, 17, 18, 19, 20, 21, 22, 23, 24, 25, 26, 27, 28, 29*" [Ellena observes attentively as Ethray allocates imaginary cards to his imaginary people seated around the table] **ETV3**	"*Check all the ones are there, 1, 11, 21… 52, so it will be the fourth person if you carry on. Because after this will be, 2, 12, 22…*" [Moves his pencil down the numbers that they used to represent people to describe the pattern] **DV4**	Ellena alerted, "*you can't just start here, you should skip two and so there is a 1, 2, 3, 4…*" **ELV3**	*Then you can see a pattern forming here (…). Wait, try and draw it like (…) I mean next to each other, so we can see on the line (…) so we can see how they form some type of pattern.* **ETV3**

MI The ability to recall/visualise an image of a formula that one has seen somewhere before or has previously learned	*... but there's supposed to be a formula*"! Ellena recalled from memory (MI) that they could find a formula to help them answer sequences related. **ELV3**	Denz suggested that they could use the formula, but neither of them could not recall the formula so, they used Figure 6.4a to derive it. **DV4**	Ellena agreed that they needed to find the area of the circle and thus started calculating this area while the discussion was still under way: $A = \pi r^2 = \pi(7)$ **ELV2**	*Because we can also use this to calculate it much faster than counting. Because now we can say, add ten to the next line, ten, ten, ten, the whole time.* ***ETV3***
KI Involves muscular activity; patterns of movement and body engagement	*Okay, so the first person gets the 1st card* [Ellena allocates imaginary cards as Ethray speaks] **ET/ELV3**	*So the first term is... he gets the first card* [compares the first term to the person getting the first card]. *The difference is, 2, 3...plus three tog...* [Moves his pencil along the written work as he looks for the common difference to use in the arithmetic sequences formula] **DV4**	*No, no, no... Remember this is standing up straight and...this...* [mimics a standing up tree with an upward movement using a pencil] **DV1**	Ethray explained to her that there was more than one way of getting the answer to the question. "*There should be a formula I know but, let's just see if we can get the answer then try to get the formula*" [overt gestures and body movements as Ethray speaks]. **ETV3**
DI Involves the processes of transforming shapes, i.e. redrawing given or initially own drawn figures	Jordan continued reading from where he paused and Denz sketched another diagram, a circle with a centre and a radius of 7 m (Figure 6.1b) DV1/JV1	Denz gestured a circular movement with his hand to show the grazing area of the goat (DI). "*So, here is the tree, you can see the tree is in the middle*" [points to the sketch as he speaks]. DV1	Jordan however argued (RPA) that as the goat moves around the tree, the rope gets shorter. He makes a sketch to illustrate his claim. JV1	In Figure 6.2, Ethray illustrated a real-life image of an actual situation in his mind (a) while Ellena enacted a mathematical transformation of the image of the same situation (b). **ET/ELV2**

KEY: ETV1 = Ethray vignette 1, **ELV2** = Ellena vignette 2, **DV3** = Denz vignette 3, **JV4** = Jordan vignette 4, etc.

was very encouraging. This is an indication that when teaching mathematics, teachers should focus both on the visual aspects of the learners' problem-solving strategies and on the algebraic. Both verbal and nonverbal thinking and reasoning are vital to problem-solving and should be encouraged among visual problem solvers.

References

Arcavi, A. (2003). The role of visual representations in the learning of mathematics. *Educational Studies in Mathematics, 52*(3), 215–241.

Begg, A. (2013). Interpreting enactivism for learning and teaching. *Education Sciences & Society, 4*(1), 81–96.

Bertram, C., & Christiansen, I. (2014). *Understanding research: An introduction to reading research.* Pretoria: Van Schaik Publishers.

Boesen, J., Lithner, J., & Palm, T. (2010). The relation between types of assessment tasks and the mathematical reasoning students use. *Educational Studies in Mathematics, 75*, 89–105. https://doi.org/10.1007/s10649-010-9242-9

Brodie, K. (2010). *Teaching mathematical reasoning in secondary school classrooms.* New York: Springer.

Brown, L. (2015). Researching as an enactivist mathematics education researcher. *ZDM - Mathematics Education, 47*, 185–196.

Cohen, L., Manion, L., & Morrison, K. (2011). *Research methods in education* (7th ed.). London: Routledge/Taylor & Francis Group.

Csíkos, C., Szitányi, J., & Kelemen, R. (2012). The effects of using drawings in developing young children's mathematical word problem solving: A design experiment with third-grade Hungarian students. *Educational Studies in Mathematics, 81*(1), 47–65.

Damiano, L. (2012). Co-emergences in life and science : A double proposal for biological emergentism. *Synthese, 185*(2), 273–294.

Dongwi, B. L. (2018). *Examining mathematical reasoning through enacted visualisation.* Unpublished doctoral thesis. Rhodes University, Grahamstown.

Dongwi, B. L., & Schäfer, M. (2019). The co-emergence of visualisation and reasoning processes in mathematical problem solving. In J. Graven, H. Venkat, E. A. A., & P. Vale (Eds.), *Proceedings of the 43rd Conference of the International Group for the Psychology of Mathematics Education (Vol 2).* Pretoria, South Africa.

Dove, I. J. (2009). Towards a theory of mathematical argument. *Foundations of Science, 14*(1–2), 137–152.

Hegarty, M., & Kozhevnikov, M. (1999). Types of visual-spatial representations and mathematical problem solving. *Journal of Educational Psychology, 91*(4), 684–689.

Hershkowitz, R., Arcavi, A., & Bruckheimer, M. (2001). Reflections on the status and nature of visual reasoning – the case of the matches. *International Journal of Mathematical Education in Science and Technology, 32*(2), 255–265. https://doi.org/10.1080/00207390010010917

Khan, S., Francis, K., & Davis, B. (2015). Accumulation of experience in a vast number of cases: Enactivism as a fit framework for the study of spatial reasoning in mathematics education. *ZDM, 47*, 269–279.

Li, Q., Clark, B., & Winchester, I. (2010). Instructional design and technology grounded in enactivism: A paradigm shift? *British Journal of Educational Technology, 41*(3), 403–419.

Lithner, J. (2000). Mathematical reasoning in school tasks. *Educational Studies in Mathematics, 41*(2), 165–190.

Lithner, J. (2008a). A research framework for creative and imitative reasoning. *Educational Studies in Mathematics, 67*(3), 255–276. https://doi.org/10.1007/s10649-007-9104-2

Lithner, J. (2008b). A research framework for creative and imitative reasoning. *Educational Studies in Mathematics, 67*(3), 255–276. https://doi.org/10.1007/sl0649-007-9104-2

Makina, A. (2010). The role of visualisation in developing critical thinking in mathematics. *Perspectives in Education, 28*(1), 24–33.

Mamolo, A., Ruttenberg, R., & Whiteley, W. (2015). Developing a network of and for geometric reasoning. *ZDM - Mathematics Education, 47*, 483–496. https://doi.org/10.1007/s11858-014-0654-3

Maturana, H. R., & Poerksen, B. (2004). *From being to doing: The origin of the biology of cognition*. Heidelberg: Carl-Auer Verlag.

Maturana, H. R., & Varela, F. G. (1998). *The Tree of Knowledge*. Boston, MA and London: Shambhala, New Science Library.

Namibia. Ministry of Education [MoE]. (2010). *The National Curriculum for Basic Education*. Okahandja: NIED.

Namibia. MoEAC. (2018). *Mathematics ordinary level syllabus*. Okahandja: NIED.

Natsheh, I., & Karsenty, R. (2014). Exploring the potential role of visual reasoning tasks among inexperienced solvers. *ZDM Mathematics Education, 46*(1), 109–122. https://doi.org/10.1007/s11858-013-0551-1

Otte, M. F., Mendonça, T. M., & de Barros, L. (2015). Generalisation is necessary or even unavoidable. *PNA, 9*(3), 143–164.

Presmeg, N. C. (1986). Visualisation in high school mathematics. *For the Learning of Mathematics, 6*(3), 42–46.

Presmeg, N. C. (2014). Contemplating visualization as an epistemological learning tool in mathematics. *ZDM - International Journal on Mathematics Education, 46*(1), 151–157.

Proulx, J. (2008). Some differences between Maturana and Varela's theory of cognition and constructivism. *Complicity: An International Journal of Complexity and Education, 5*(1), 11–26.

Prusak, N., Hershkowitz, R., & Schwarz, B. B. (2012). From visual reasoning to logical necessity through argumentative design. *Educational Studies in Mathematics, 79*(1), 19–40.

QSR International. (n.d.). Retrieved April 4, 2019, from https://www.qsrinternational.com/

Rivera, F. D., Steinbring, H., & Arcavi, A. (2014). Visualization as an epistemological learning tool: An introduction. *ZDM – International Journal on Mathematics Education, 46*, 1, 2.

Simmt, E. S. M., & Kieren, T. (2015). Three "moves" in enactivist research: A reflection. *ZDM – Mathematics Education, 47*(2), 307–317.

Sochański, M. (2018). What is diagrammatic reasoning in mathematics? *Logic and Logical Philosophy, 27*, 567–581. https://doi.org/10.12775/llp.2018.001

Staples, M. E., Bartlo, J., & Thanheiser, E. (2012). Justification as a teaching and learning practice: Its (potential) multifacted role in middle grades mathematics classrooms. *Journal of Mathematical Behavior, 31*(4), 447–462.

Swafford, J. O., & Langrall, C. W. (2000). Grade 6 students' preinstructional use of equations to describe and represent problem situations. *Journal for Research in Mathematics Education, 31*(1), 89–112.

Tripathi, P. N. (2008). Developing mathematical understanding through multiple representations. *Mathematics Teaching in the Middle School, 13*(8), 438–445.

Van Garderen, D., Scheuermann, A., & Poch, A. (2014). Challenges students identified with a learning disability and as high-achieving experience when using diagrams as a visualization tool to solve mathematics word problems. *ZDM – International Journal on Mathematics Education, 46*(1), 135–149.

Webb, N. M. (1991). Task-related verbal interaction and mathematics learning in small groups. *Journal for Research in Mathematics Education, 22*(5), 366–389.

7 Visualisation and Algebra

From Formulas to Patterns

Sindisiwe Herbert

Introduction

Elementary algebra is typically considered an essential and also an abstract, non-visual part of any Grade 8 or 9 learner's experiences with mathematics. It allows learners access to complex ideas across mathematical disciplines, through fluency in the 'language' of variable. However, algebraic thinking is a skill that transcends variable manipulation and is often introduced with pictorial patterns. The study described in this chapter was designed to explore the outcomes of an activity that reversed this order and required learners to start with an algebraic expression and invent a visual pattern to represent it. Both the activity and the context in which it took place were intentionally open-ended, low-tech and creative, which meant that the resulting ideas would be accessible to anyone. The aim was to bring a highly visual and kinaesthetic focus to the topic of algebraic expressions in order to broaden and deepen learners' understanding of the underlying concepts.

Context

Algebra is the written language of mathematics, with a syntax and structure that allows mathematicians to communicate the mechanics of a calculation with speed and accuracy. Consider the theorem which states that *the square of the hypotenuse of a right-angled triangle is equal to the sum of the squares of the two shorter sides*, and the speed at which that statement can be encoded as $a^2 + b^2 = c^2$. For this reason alone, it would be necessary for students to learn the conventions of algebraic manipulation along with (and often before) the theories of geometry, trigonometry and calculus. However, algebraic *manipulation*, like the variables that make it what it is, represents something much greater than itself. The real ideas of algebra are, indeed, much older than the encoding system with which I have been synonymising them. As Radford (2008) points out, "Chinese mathematicians thought in algebraic ways without using letters and Euclid used letters without thinking algebraically" (2008, p. 84). The word 'algebra' itself can be traced to "the title of a book, '*ilm al-jabr wa'l-muqābala*', 'the science of restoring what is missing and equating like with like', by the mathematician al-Ḵwārizmī" (Oxford

DOI: 10.4324/9781003172420-9

Dictionary, 2020). This book, written without modern variables and syntax, was an important step in mathematicians' long, continuous quest to get beneath the specifics of quantity and discover the relationships underlying all numbers – that is, to *generalise* arithmetic. The pursuit of generalisation has led mathematicians to *abstract* algebra – the invention and study of structures such as groups, rings and fields – but throughout this chapter the term 'algebra' is used in the sense of *elementary* algebra, as learned at a high school level.

Generalisation, says Radford, "means to select some features of *a* and *b* and to dismiss some others" and is a necessary skill, not only in mathematics but in the creation of any "concept ... that does not fully coincide with any of its instances" (Radford, 2008, p. 83). In the context of algebra, the ability to find an *input-output* relationship between two sets of numbers is considered an essential form of generalisation, often referred to as algebraic reasoning.

Algebraic reasoning is a critical skill in mathematics, often introduced through the study of the linear patterns known as arithmetic sequences. Radford (2013), Pegg & Redden (1990) and English & Warren (1998), among others, describe patterning activities designed to lead students to a description of the general term of a sequence. These sequences, being a well-recognised entry point into algebraic thinking and a notable part of the South African curriculum (Department of Basic Education, 2011), were chosen as the vehicle for a study into visualising algebra. However, unlike most previous studies and typical classroom resources such as textbooks, the activity central to the study was designed to start from an algebraic expression and require participants (Grade 9 learners with some pre-existing experience with algebra) to create a visual sequence that represented it. There is good reason for starting with a concrete, visual representation and moving to a more abstract, algebraic expression, since the algebra is, by design, less specific and more readily applicable to other contexts. However, there is an assumption that "learning mathematics begins with the concrete and 'ascends' (the metaphor is pertinent) to the abstract" (Noss et al., 1997, p. 204), an assumption that is worth questioning. The process of finding an algebraic expression to represent a visual sequence is well-known and easily algorithmised. Indeed, many learners deal with these problems by counting the number of objects in each picture and working purely numerically, completely disengaging the problem from its intended context and purpose (Samson, 2007, p. 25). The reverse process is far more open-ended, with an infinite number of possible pictorial representations for any given algebraic expression. Engagement with visual and algebraic representations in both 'directions' is also useful practice in changing register – a critical skill in mathematics.

The study in this chapter was inspired heavily by a DiSessa et al. (1991) article that describes the process by which a group of Grade 6 learners "invented graphing as a means of representing motion" (DiSessa et al., 1991, p. 117). In a small group, learners created, discussed and refined their representations, gaining not only fluency in graphing but also an ability to critique and improve on representations of all kinds, a skill that DiSessa et al. call "meta-representational competence" (1991, p. 117). The activity described in their paper is strongly constructivist in orientation, which lent a constructivist bent to the study discussed here. However, constructivism falls short of capturing the dual nature of mathematics as a system of logical relationships that exists not only within the mind of the mathematician but also in a collective, social and historical space. Radford's concept of knowledge objectification builds on this sociocultural concept, describing knowledge as "a dynamic and evolving implicit or explicit culturally codified way of doing, thinking, and relating to others and the world" (2015, p. 550). In this framework, learning is not a process of creating and refining your own systems of interacting with and interpreting the world around you. Rather, it is a series of experiences through which you are gradually made aware (often by other people) of things that had previously remained obscure to you, *realising* a concept in the literal sense of being able to "treat a general structural relationship as an object in its own right" (Presmeg, 2016, p. 22). Radford calls these experiences "processes of objectification" (2015, p. 559). Tied up in these processes of objectification are "processes of subjectification" because engagement with knowledge means an engagement with the society that embodies that knowledge, and through your learning you take a position as a part of that culture (Radford, 2015). In the context of this theory, it is important to take note of *how* learners engage with an activity, particularly in moments when they *notice* elements of the subject at hand, showing their conscious attention through gesture, questions, explanations and the like.

The theory of knowledge objectification is particularly relevant in mathematics in general, and algebra in particular – given that mathematical objects are almost entirely abstract in nature and cannot be accessed without the use of some intermediary symbol or diagram. For example, the number four can be referred to with the numeral '4' or a collection of four dots, but the number (and possibly the symbol itself) is purely conceptual. The study of the use of these intermediaries and their relationships to the concepts themselves is semiotics. Presmeg describes the concept as the "object", the intermediary representation as the "sign vehicle" and the relationship between the two as the "sign" (2006, p. 21). Often, as in the example of the number four, there are multiple ways of representing a mathematical object with a sign and these signs relate to each other in multiple ways. Duval refers to these semiotic systems as "registers of representation" (1999, p. 6), each of which can foster a different kind of engagement with the object. Changing between these different signs was referred to earlier as *converting between registers* and was an important part of this study's design. When taking part in the patterning activity, learners were expected to engage with and switch between various signs and registers. Notably, some of these were historical, culturally recognised signs like '$3x + 1$' or 1;4;7;10;13;...' while others were novel, idiosyncratic signs – the visual representations created by themselves and other participants. DiSessa et al. describe the feeling of "ownership of ideas and artifacts" (1991, p. 124) as a strong motivator for interest and thus engagement in an activity. This kind of imaginative engagement with mathematics is important and rare. However, it is equally important for learners of mathematics to be able to communicate effectively, and to do so they must have a level of competence in dealing with conventional mathematical signs. This is part of what makes fluent register switching so important. Moreover, Presmeg notes that "conversions amongst registers may facilitate ... *objectification* in mathematics" (2006, p. 22) – that is, engaging with translations between different representations of a concept is a large part of understanding the concept itself.

The second source of inspiration for the study was the research of Stott & Graven (2013), who ran informal after-school mathematics clubs with Grade 3 learners, which "focused on developing a supportive learning community where learners can develop their mathematical proficiency, make sense of their mathematics and where they can engage and participate actively in mathematical activities" (Stott & Graven, 2013, p. 2). This was in response to the environment of many South African classrooms, in which the pressure to complete the curriculum and perform well in assessments often leads to anxiety, disengagement and a culture of "teaching for/to assessments" (Stott & Graven, 2013, p. 2). The community of enquiry that these clubs are intended to create is just as important in a high school setting as in a primary school setting, and this led to the creation of a Grade 9 mathematics club at a public boys' school in South Africa, in the context of which the research took place. The club consisted of around 12 regular attendees, along with several learners who attended only occasionally, and two facilitators, one of whom was a current teacher at the school and the other (myself) a former teacher. The structure of the club sessions was based on Radford's four stages in classroom activities (2013) and Liljedahl's Thinking Classroom model (2016).

The final decision in designing the study was choosing the materials with which to create these visual representations. A digital application such as *Desmos* or *Geogebra* would have had the advantage of being programmable so that a single representation could be manipulated to depict many different expressions. However, in practice, it was difficult to ensure that all participants had access to the necessary technology or data to use these apps. The school at which the research was carried out had two computer laboratories, but one was always in use after school and the other was full of slow and occasionally unreliable hardware. This is typical of many South African schools – learners have smartphones but no access to the Wi-Fi or data necessary to download visual mathematics apps, and where there are computer labs, often they are either unreliable or busy to the point of impracticality in lesson planning. For this reason, a physical building material was decided on. The original idea was to collect and use bottle caps, but a pilot study was carried out using colourful building cubes and this worked so well that it was adopted for the main study. The different colours, linking abilities and three-dimensionality of the cubes allowed for interesting interpretations of the brief that would not have been possible with bottle caps. However, the general ideas of the activity could be carried out in any medium, from a digital modelling programme to a collection of coloured stones or simple pencil and paper.

The Study

In the setting of the aforementioned after-school maths club, learners were asked to choose a partner and spend four club sessions (which took place once a week) creating visual representations of linear algebraic expressions with the medium of colourful building cubes. A few weeks after the final club session dealing with the activity, each pair of learners took part in a task-based interview, during which they designed one more representation and discussed their process and inspiration for previous creations.

Having learned about both algebraic expressions and linear sequences in Grade 8, the learners already had an introduction to the subject and were fluent in substituting numbers for variables, a necessary prerequisite for the activity. They had also been attending club sessions for several months, so they were used to the exploratory nature of the club content and had begun to develop a community in which their ideas were regularly centred and discussed, rather than a community that revolved around a teacher.

During the study, I hoped that learners would develop a deeper understanding of both algebraic expressions and linear sequences. According to Radford, "forms of algebraically ... perceiving and dealing with sequences are codified forms of thinking and doing ... refined in human cultural history" (2013, p. 15). Through this activity, the club members invented new ways of codifying algebraic expressions and linear sequences, while reflecting on well-established conventions and patterns of thought that would serve them well in their cultural position as learners and practisers of mathematics.

Jeff and Dev

Jeff and Dev (referred to here by pseudonyms) were dedicated members of the Maths Club from early on. They participated actively in each session's activity and expressed enthusiasm for learning mathematics beyond the scope of the Grade 9 syllabus. During the study, they spent each session expanding on the previous week's representation, which meant that the development of their ideas over the course of the study was clear and their logic easy to follow. They were also frontrunners: always among the first to come up with ideas that became popular in the group as a whole. It was no surprise that they chose to work together for the study since they were friends and made a good team. Jeff was confident and quick to offer insights, while Dev was thoughtful and keen to understand, often asking questions that would spark a new line of enquiry.

In the first club session of the study, the learners were asked to create a visual representation of the expression $2x$. Jeff and Dev's first representation was common to many pairs at the beginning of the study and looked like a set of 'towers'. In this representation, each column represents a term and they are arranged in ascending order. Figure 7.1 shows Jeff and Dev's white and black representation. They switched from white to black when they ran out of white blocks.

In Session Two, to represent the expression $3x + 1$, Jeff and Dev improved on their towers, changing their use of colour to differentiate between terms as they had seen many other pairs do in the first session. Dev had the idea of including grey towers below the main towers to indicate the value of x for each term. This proved to be pivotal for the development of what I called the 'Visual

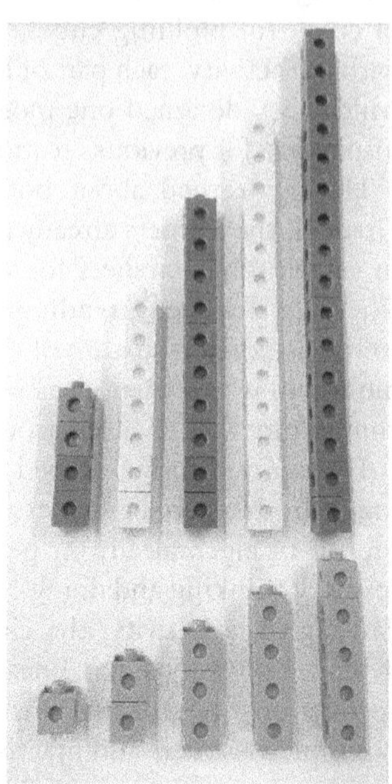

Figure 7.1 Jeff and Dev's white and black towers from Session One (left) and updated towers from Session Two (right).

Expression' representation that they created in Session Three. This third iteration was still an expansion of the original towers idea, although by now it had begun to look very different. Jeff extended Dev's idea of including an indication of the value of x, by introducing blocks in different colours to represent each part of the expression. Figure 7.2 shows their representation of the expression $2x+3$, in which, to the left of the mid grey blocks, the number of dark grey blocks shows the value of x for the term, the three light grey blocks represent the addition of three and the two black blocks represent multiplication by two. They began by building only the column on the left, but Jeff then added an '*equals sign*' in the form of the two mid grey blocks, and they included the value of the term, or as they called it, '*the answer*' on the right. They continued to show the additional three blocks with light grey blocks and used dark grey blocks to represent the value of $2x$ for each term.

In T_1 and T_2, Jeff placed the dark grey 'answer' blocks in a column. However, in T_8 he built these blocks into a 2×8 array. When I asked him why, he explained that it was just a space-saving measure since a column of 16 blocks would be inconveniently tall. I pointed out that in T_8 the x-column and the answer rectangle had the same height (8 blocks) and that this was perhaps useful, as it showed the multiplication of $2 \times x$ in the width and height of the rectangle. We then modified T_1 and T_2 to match T_8, as shown in the final image in Figure 7.2.

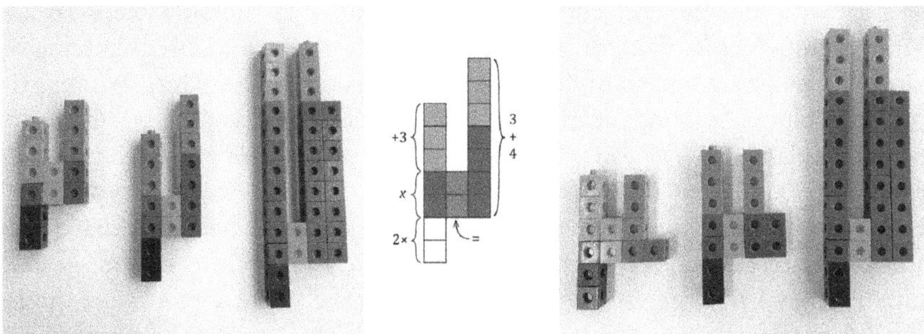

Figure 7.2 Jeff and Dev's first Visual Expression showing T_1, T_2 and T_8 (L), an explanation of T_2 (middle) and their modified Visual Expression (R).

Figure 7.3 Jeff and Dev's second Visual Expression before (left) and after (right) modifications.

Jeff and Dev built another Visual Expression in Session Four to represent $3x - 2$, using white blocks to show x, orange blocks to show 'multiplied by three' and pale white blocks to show 'subtract two', as shown in Figure 7.3. Because I had encouraged them to build rectangles with a base of two blocks in Session Three, they put their black 'output' blocks into two columns. In the interview, Jeff stated that it was "*easier if... we had two rows because we'd be counting upwards in twos, not in ones*". I explained again how a rectangle with base three might show multiplication by three and they modified their representation to show x on the left and 3 at the bottom with a $3 \times x$ rectangle of black blocks showing the answer, two of which were replaced with pale white ones to show the subtraction.

In their interview, conducted two weeks after the fourth club session, Jeff and Dev were asked to build a representation of $4x - 3$. As soon as they started discussing what they would build, Dev asked, "*So what colour's going to represent x?*" It was clear that they had become attached to the idea of the Visual Expression, as they continued to discuss what colour would represent "*minus three*" and "*the answer*" without explicitly discussing that they were going to build this kind of representation. This had become *the* representation and they both had the concept clear enough in their minds that the only necessary decisions were which colours to use and how to lay them out. After agreeing on colours, they proceeded to build a Visual Expression of pale white, white, grey and black blocks running left to right that Jeff described as "*straightforward*". As Dev explained, "*The white represents the term, the pale white represents the coefficient of x, the grey represents the minus three, then the black is your product*". (Dev often misused the term *product*, using it to refer to what would have been more accurately described as the *output* or, as Jeff referred to it, the *answer*.) Initially, they had organised the sequence with the white x blocks on the far left, before the pale white 'multiplied by four' blocks. Jeff then swapped them, saying, "*Wouldn't it be easier if we actually did this? 'cause now you can say four x minus 3*". We discussed that it was not important mathematically, because, as Dev pointed out, "*Order doesn't matter in multiplication*". But Jeff insisted that it was better to have the visualisation conform to the conventions of algebraic expressions, in which multiplication is written with the constant coefficient before the variable/s. This was a clear indication of what had happened in the creation of the Visual Expression, not just between Jeff and Dev, but among most of their peers as well. As Jeff described it, "*It's like writing the equation down and you can see it – exactly what to put in the equation and the answer*". In attempting to build the best possible visual representation of an algebraic expression, learners found the most efficient solution – a 'word-for-word' translation from symbols to colours.

There were several very interesting moments of Jeff and Dev *noticing* and *failing to notice* during the interview. For example, Dev noticed that the outputs of $4x - 3$ twice included perfect squares ($T_1 = 1$ and $T_3 = 9$) and made the false assumption that all of the output values would be perfect squares. He stated that a different expression would not form part of the same sequence "*because [they are all] perfect squares*". He quickly realised that this was incorrect when I pointed out that $T_2 = 5$. Later, he noticed something that he was ostensibly taught in Grade 8, which he displayed by asking, "*In every linear sequence … is the difference of your products going to be the coefficient of x every time?*" We discussed why this was the case and while Jeff and I moved on to discuss why they chose to put their blocks into tall towers, Dev continued to consider this idea, later interrupting to ask, "*So on a graph, ma'am, it would be each time your x increases, your [output] increases by two?*" Through engagement with this activity, he *objectified* this feature of linear numerical relationships and started applying it not only to sequences but also to graphs. This opened a conversation about the similarities and differences between linear sequences and linear graphs.

Jeff noticed that putting blocks into multiple rows makes them "*easier to count*", but despite our conversations about it in Sessions Three and Four, neither Jeff nor Dev fully grasped the fact that a $d \times x$ array can be used as a shortcut for multiplication. Jeff noticed obliquely that an array shows divisibility, saying "*In this term … if we had two rows then there'd be one extra*". Later, Dev expressed the same idea more explicitly: "*If your [outputs] are even numbers, then you can have two rows*". The definition of even numbers being multiples of two means that this certainly shows the beginning of an engagement with this concept, but it was not one we were able to fully explore during the interview.

In the end, Dev said that their third representation was "*confusing*" and Jeff maintained that the fourth representation was "*very hard to understand*". He explained that, in the fourth representation, "*everything's all over. Here [in the interview representation] it's moving left to right straight*". Dev agreed that their final representation was the best and clearest of their designs, confirming that they had become focused on visually recreating the algebraic expression as nearly as possible. Despite our discussions on the topic, they needed more time with the subject to expand their observations on even numbers to outputs divisible by any natural number.

Other pairs

Of the six pairs involved in the study (all referred to by pseudonyms), Jeff & Dev showed the clearest progression of ideas from session to session. Kylian & Warren and Sam & Loyiso's work was disrupted by their sporadic absenteeism. Mike & Deon's disagreements on how the prompt should be answered led to them working separately for the first two sessions. Philip & Kabelo liked their first representation so much that they struggled to change it, even when directly encouraged to do so. Conversely, Ranveer & Alex insisted on creating something new each week, until they were encouraged to revisit their first attempt. Each pair created unique representations along the way, and all except for Ranveer & Alex ended up with some form of Visual Expression in the interviews. Here Sam & Loyiso, Philip & Kabelo and Ranveer & Alex's representations will be reviewed in order to provide a comparison to Jeff & Dev. Where necessary, drawings will be used rather than photographs in order to concisely show the meanings of some of the more obscure representations.

In the first session, almost all the pairs created a collection of rows, columns or other groups representing $2;4;6;8;\ldots = 2x$. Some were arranged in towers like Jeff & Dev, and others in pyramid shapes, spirals and rectangles. Ranveer & Alex first built a set of towers like Jeff & Dev, but upon seeing that several other pairs had done the same, rearranged their groups into $2 \times x$ rectangles.

In the second session, to represent $4;7;10;13;\ldots = 3x+1$ (Figure 7.4), the participants continued to focus (as expected) on the sequence outputs in the second session – only Jeff & Dev thought about representing any other part of the expression at this point. Philip & Kabelo edited their previous representation, improving on their use of colour so that a change in colour indicated a new term, which Sam & Loyiso had done in Session One. Loyiso was absent, so Sam created a spiral that turned into a kind of zig-zag. Colour changes indicated new terms and the flow is indicated on the diagram with arrows. Ranveer & Alex built a set of nesting 'L' shapes, with each larger 'L' representing the next output of the sequence.

In the third session (Figure 7.5), the participants were actively encouraged to create a representation in which looking at the eighth term alone would

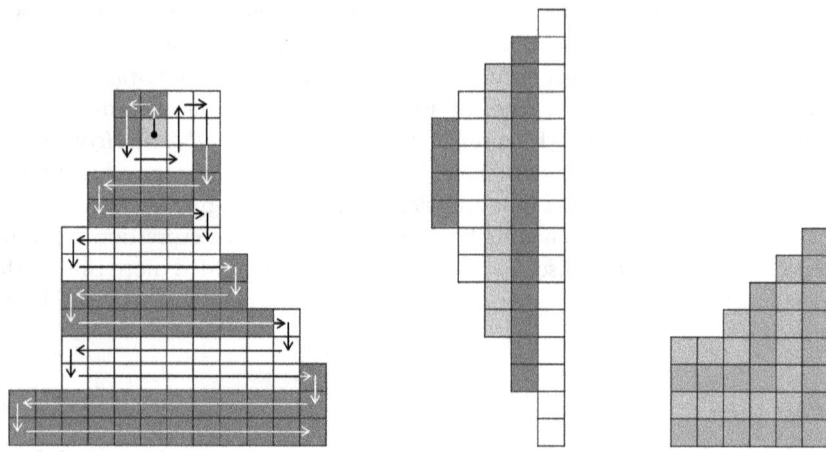

Figure 7.4 (Left to right) the second session representations of Sam, Philip & Kabelo and Alex & Ranveer.

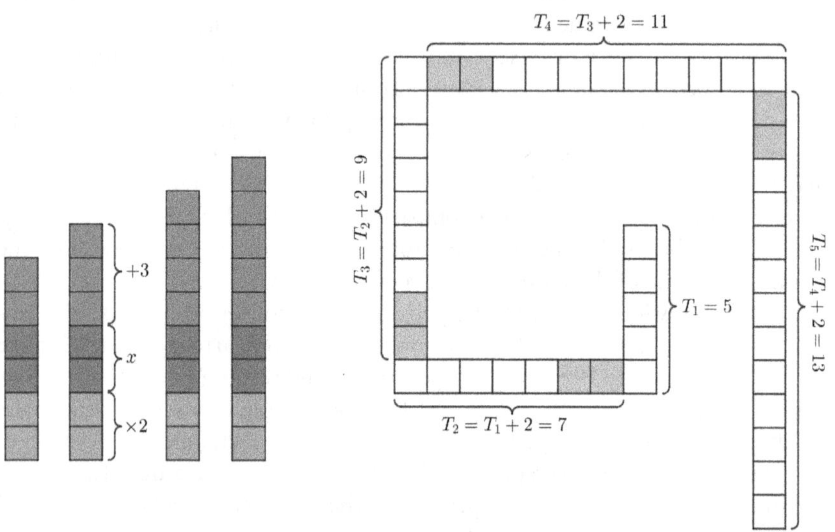

Figure 7.5 (Left to right) the third session representations of Sam & Loyiso and Alex & Ranveer.

give more information than merely the output value. The expression was $5; 7; 9; 11; \ldots = 2x + 3$. Sam & Loyiso abandoned the output values of the sequence, creating alongside Jeff and Dev one of the first Visual Expressions. Alex & Ranveer continued to make something new in each session (as was their habit) and built a spiral, adding two new orange blocks to each term of the

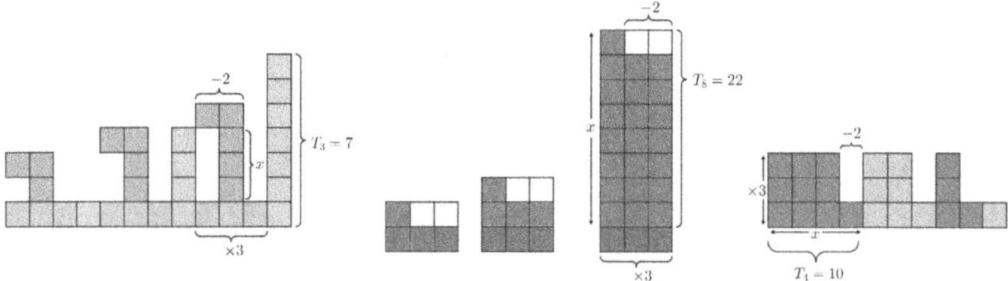

Figure 7.6 (Left to right) the fourth session representations of Sam & Loyiso, Philip & Kabelo and Alex & Ranveer, with explanations beneath.

sequence. Philip & Kabelo built a pyramid again and when I asked whether the eighth term on its own showed any details about the expression, they became discouraged and broke their representation apart without enough time to create something new. After a quick whole-club debriefing they drew Jeff & Dev's representation in their journals.

Session Four (Figure 7.6) saw most participants adopting some form of Visual Expression to represent 1; 4; 7; 10; … = $3x - 2$. With my prompting, Alex & Ranveer returned to their rectangles from the first session and ended up with arrays very similar to Jeff & Dev's. After their frustration in Session Three I gave Philip & Kabelo some guidance and they also ended up with an array. Sam & Loyiso stayed with their Visual Expression, adding a column to represent the output value of each term.

The interviews were very revealing. I asked each pair to "think out aloud" as they created a representation of 1; 5; 9; 13; … = $4x - 3$. Mike & Deon, who had often produced two very different representations during the club sessions, seemed to have come to an agreement and produced only one. However, Sam & Loyiso, Kylian & Warren and Alex & Ranveer all came up with separate representations, diverging at various points during the building process. Philip & Kabelo and Alex & Ranveer noticeably tried to recreate their representations from the fourth club session, but because the interviews took place around two weeks after the final research session, they had forgotten exactly what they had done before. The elements that they remembered provided insight into what they had *noticed* about their representations.

Sam & Loyiso produced two Visual Expressions very similar to one another. All of their explanations and suggestions for expanding on their representations showed evidence of a desire to replicate the visual mechanics of algebra in block form. Like Jeff, they insisted on using the conventional algebraic order of coefficient, variable and then the constant term. When I asked them how they might represent x^2, Loyiso put two white blocks (the 'square') on top of a column of dark grey blocks (the 'x'). Sam purposefully used a horizontal row to represent the −3, explaining that "*since the dark grey is horizontal it's a minus sign*". He was initially thinking of subtraction rather than addition of a negative, explaining that if the constant was positive he would arrange the blocks into a + sign. However, as we moved on to discuss how we could represent an expression with a negative coefficient, Sam decided that it was more efficient to use horizontal rows to represent negative numbers and vertical columns to represent positive numbers, which was a very useful idea that did not come up in any other discussions. Sam had other, similarly inventive ideas, such as condensing the very tall answer column with different colours, using factors such as "***one block could represent a certain number***".

Philip and Kabelo chose the same colours for their interview representation as they had used in the fourth club session, intending to recreate it. They started with groups of blocks representing the output values of each term and then organised these into individual arrays with added blocks to show the multiplication and subtraction. In the interview, however, they began by building a representation that, much like Loyiso's, did not include any indication of the output value. Their focus, in trying to recreate their previous representation, was on which colours they had used to represent different parts of the expression, showing that what they had noticed and remembered was the Visual Expression aspect rather than the array. When they realised that, contrary to their expectations, the total number of blocks in each term was not equal to the term output, they reorganised the terms and added a column to represent the term output, conforming to what had become the standard representation.

Ranveer & Alex also intended to recreate their representation from the fourth session during the interview, but were the only pair not to settle on a Visual Expression. This reflected the desire that they displayed right from the first session, to create something unique. Alex built two-columned models of the output of T_1, T_2, T_3 and T_4, intending to show '-3' with the difference between the columns. Ranveer created a similar two-columned structure representing the outputs of the equation, but instead, he noticed the fact that (coincidentally) the smaller column in T_3 was three blocks high, corresponding with the value of x. Interestingly, both of these features were present in their previous representation, but on this occasion, with only two rather than x columns, they were able to only show either the subtraction of 3 or the value of x, rather than incorporating both of these aspects of the equation into their visualisation. Ranveer was the only participant to use the third dimension to show something mathematical about the sequence. He stacked the terms behind each other in order to show the constant difference of four between each term and the next.

Conclusions and Discussion

Ultimately, most of the participants seem to have gained extra fluency in register switching – an idea of the power in algebra's simplicity, and various other insights – along with an increased belief in their ability to do mathematics. Over the course of the study, the participants developed some wonderfully creative ideas for representing algebraic expressions visually, but unlike the learners in DiSessa et al.'s study (1991) they did not produce a product that coincided with either common mathematical methods or my idea of the ideal representation. This was not a problem, since the activity nevertheless required them to engage with the culturally recognised forms of representing linear functions as an expression or a sequence. However, it highlights the insufficiency of constructivism as a universal theory of learning. To come to an increasingly comprehensive understanding of linear algebraic expressions, the participants cannot merely be presented with ideas and activities in order to 'build' knowledge from their own experiences (Radford, 2013, p. 8). Without continued engagement with the mathematical community in the form of teachers, classmates, textbooks and the like – particularly in the context of joint activity – there will always be things that they fail to "notice" (Radford, 2013, p. 24).

It would be interesting to see the concept of creating visual representations of algebraic expressions adapted for use in a more typical South African classroom setting with more learners, fewer resources and less time than was available here. Would the activity stimulate similarly enthusiastic engagement in a situation where participation was not voluntary? Could it have a positive effect on the mathematical beliefs and attitudes of learners who were not already interested in mathematics beyond the bounds of the syllabus? There are other questions raised by the study, too. Are after-school clubs at a high school level as effective a space for creative mathematical activity as this study suggests? Would these kinds of visual representations be useful in further algebraic studies – not just representing, but perhaps manipulating algebraic expressions?

If, as Radford claims, "how we come to know is shaped by, and consubstantial with, the activity through which knowledge is instantiated" (2013, p. 38) then it is critical that we continue to research and design more effective activities for the purposes of teaching and learning mathematics. I believe that, particularly in a Southern African context, the patterning activity that formed the basis of this study can help to mitigate issues caused by minimal resources, learners with low mathematical confidence and overly teacher-centred classrooms. This is because the activity requires few materials and puts ownership of knowledge and responsibility for learning into the students' hands. Changing the teacher from the ultimate source of knowledge in the classroom to just one of many access points into the mathematical community increases learners' belief in their own ability as mathematicians, their willingness to engage actively in the task, and thus their proficiency in the necessary mathematical concepts and means of communication. It also, perhaps most evidently here, encourages creativity and thus different perspectives on the topic of algebra, giving participants the opportunity to learn from each other. The activity showed itself to be an effective and exciting way to provide learners with tangible (visual and kinaesthetic) access to algebraic concepts that are at times obscure.

References

Department of Basic Education. (2011). *National Curriculum Statement: Curriculum and Assessment Policy Statement Mathematics Senior Phase Grades 7–9*. Pretoria: Government Printing Works.

DiSessa, A. A., Hammer, D., Sherin, B., & Kolpakowski, T. (1991). Inventing Graphing: Meta-Representational Expertise in Children. *Journal of Mathematical Behaviour, 10*, 117–160.

Duval, R. (1999). Representation, Vision and Visualization: Cognitive Functions in Mathematical Thinking. Basic Issues for Learning. *21st Annual Meeting of the North American Chapter of the International Group for the Psychology of Mathematics Education. 1*, pp. 3–26. Cuernavaca, Mexico: PME-NA.

English, L., & Warren, E. (1998). Introducing the Variable Through Pattern Exploration. *The Mathematics Teacher, 91*(2), 166–170.

Liljedahl, P. (2016). Building Thinking Classrooms: Conditions for Problem-Solving. In P. Liljedahl, *Posing and Solving Mathematical Problems* (pp. 361–386). Cham: Springer.

Noss, R., Healy, L., & Hoyles, C. (1997). The Construction of Mathematical Meanings: Connecting the Visual with the Symbolic. *Educational Studies in Mathematics, 33*(2), 203–233.

Oxford Dictionary. (2020, 03 14). *Algebra*. Retrieved from Lexico: https://www.lexico.com/definition/algebra

Pegg, J., & Redden, E. (1990). Procedures for, and Experiences in, Introducing Algebra in New South Wales. *The Mathematics Teacher, 83*(5), 386–391.

Presmeg, N. (2006). A Semiotic View of the Role of Imagery and Inscriptions in Mathematics Teaching and Learning. *30th Conference of the International Group for the Psychology of Mathematics Education, 1* (pp. 19–34).

Presmeg, N. (2016). Semiotics in Theory and Practice in Mathematics Education. In N. Presmeg, L. Radford, W.-M. Roth, & G. Kadunz, *Semiotics in Mathematics Education* (pp. 5–29). Cham: Springer.

Radford, L. (2008). Iconicity and Contraction: A Semiotic Investigation of Forms of Algebraic Generalizations of Patterns in Different Contexts. *ZDM, 40*(1), 83–96.

Radford, L. (2013). Three Key Concepts of the Theory of Objectification: Knowledge, Knowing and Learning. *Journal of Research in Mathematics Education, 2*(1), 7–44.

Radford, L. (2015). Methodological Aspects of the Theory of Objectification. *Perspectivas da Educação Matemática, 8*(18), 547–567.

Samson, D. (2007). *An Analysis of the Influence of Question Design on Pupils' Approaches to Number Pattern Generalisation Tasks*. Unpublished MEd thesis. Grahamstown: Rhodes University.

Stott, D., & Graven, M. (2013). The Dialectical Relationship between Theory and Practice in the Design of an After-school Mathematics Club. *Pythagoras, 34*(1), 1–10.

Part 3

Visualisation and Technology

8 Learning with *GeoGebra*

Deepak Mavani

Introduction

At the heart of this chapter is an investigation on how learners learn mathematical concepts through the medium of GeoGebra, in the context of a teacher intervention programme called the GeoGebra Literacy Initiative Project (GLIP). GeoGebra is a freely available dynamic geometry software (DGS) package and GLIP is a teacher development initiative in Mthatha in the Eastern Cape of South Africa. GLIP aims to grow and develop appropriate Information and Communication Technology (ICT) skills in teachers, to harness the teaching and learning-specific potential of GeoGebra. GLIP also constitutes the empirical site of a PhD research project, out of which this chapter is crafted. The GLIP programme was designed in numerous cycles, each of which was driven by a specific mathematical idea or topic. This chapter focuses specifically on the first cycle of GLIP with the topic *the angle at the centre of a circle.* Mavani et al. (2018), reporting on this cycle, discuss how selected GLIP teachers collaboratively developed lesson resources in the form of applets. By incorporating co-developed applets into their routine classroom practice, participant teachers brought forth GeoGebra as an effective learning and visualisation tool in their classrooms. This chapter reports on the first cycle of GLIP from the perspective of the learners, as their teachers implemented these co-developed applets.

Epistemological Access and Technology

Much research has been conducted regarding the use of technology (technology and ICT are used interchangeably in this chapter) in South Africa in learning and teaching mathematics (De Villiers, 1996; Mudaly, 2010; Stols & Kriek, 2011). However, there is only limited evidence of research available about how technology has been taken up in previously disadvantaged communities such as in the Eastern Cape (Mavani et al., 2018). Learners in developing regions such as Mthatha in the Eastern Cape (capital of the former Transkei), dominated by previously deprived communities, historically had (and still have) limited epistemological access to mathematics education. This is due to poor education systems of the erstwhile apartheid homelands (of which the Transkei was one)

DOI: 10.4324/9781003172420-11

with their ill-qualified and inadequately supported teachers. This led to gross inequalities in terms of both access to and engagement with quality education. It was only since the beginning of the new democratic era in 1994 that the national Department of Basic Education in the new South Africa took the initiative of incorporating technology into teaching and learning contexts across the entire region (Stols et al., 2008). Despite this initiative, however, the Ministerial Committee (2013) formed by the Department of Basic Education reports that where ICT resources are available in schools, they are still not used to their full potential. It is against this backdrop that this chapter, in the context of my bigger PhD study, hopes to address this gap in schools in Mthatha in the Eastern Cape.

Research Question

The scholarly works of Arcavi (2003), Presmeg (1986, 2014), Duval (1999, 2013) and other researchers have provided compelling empirical evidence of the important role of visualisation in developing an understanding of mathematical concepts and knowledge. Their research found strong evidence that learners do employ different visual strategies to construct meaningful conceptual ideas. In particular, Presmeg (1986), commending the use of visual imagery, claims that the embodiment of abstract ideas in a concrete image can be effective in learning mathematics, i.e. the use of visualisation can facilitate the access to and learning of mathematical content with understanding.

Research also recognises the contributing potential of technology in visualisation (Arcavi 2003; Stols & Kriek, 2011). An important aspect of technology is that when used appropriately, through the power of its inherent visual elements, it can enhance learning so that learners understand the mathematical concepts presented (Mudaly, 2010). This chapter specifically focuses on how the visualisation capacity of *GeoGebra* enabled the exploration and learning of the *angle at the centre of a circle* theorem in Euclidean geometry, by specifically asking the question: **What visualisation role can GeoGebra play in the learning of circle geometry?**

Visualisation in Mathematics

Hilbert (1952) asserts that "[w]ith the aid of visual imagination, we can illuminate the manifold facts and problems of geometry". Hilbert's statement is significant in the learning of mathematics, where visualisation has a powerful role to play in supporting and illustrating symbolic results (Arcavi, 2003). Very importantly, Arcavi argues that "[v]isualization is no longer related to the illustrative purposes only but is also being recognized as a key component of reasoning (deeply engaging with the conceptual and not the merely perceptual), problem solving, and even proving" (p. 235). Building on from research across the globe, Arcavi (2003) further suggests that visualisation consists of processes that lead "to the development and use of an intervening conceptual structure which enables us to see through the same visual display, things similar to those seen by an expert" (p. 234).

The effectiveness of visual processing in learning mathematics is also recorded by other researchers such as Presmeg (1986, 2014), De Guzmán (2002) and Duval (2013). Presmeg (1986) isolates different kinds of visual imagery (especially pattern imagery and dynamic imagery) that play an important role in learning mathematics with understanding, and are even suited to the generalisation of concepts. In a research report, Presmeg (1991) suggests that teaching and learning that promotes the effective use of visualisation is all about making *connections*, "[t]he visual teachers constantly made connections between the subject matter and other areas of thought, ... and above all, the real world" (p. 194). For example, when drawing a rectangle on a piece of paper, or cutting it out, the visualisation process includes making corresponding connections to its properties, and also to real-life situations or concrete objects.

De Guzmán (2002, para. 2) construed mathematical visualisation as:

> ...a way of acting with explicit attention to the possible concrete representations of the objects that one is manipulating in order to have a more efficient approach to the abstract relationships one is handling.

Visualisation is an important aspect in mathematical activity through which learners can explore different structures of concrete reality. Concrete manipulations and visual approaches very often give learners intuitions to conjecture abstract and symbolic objects. The basic mathematical ideas are borne from concrete and visualisable situations, and therefore an important aspect of visualisation in mathematics lies in generalisations. For De Guzmán (2002), visualisation is a dynamic process that is difficult to replace with the written word.

Interestingly, in their study of learners' responses to a geometric task, Mhlolo and Schäfer (2013) observed that "the learners were making connections based on their subjective impressions" (p. 86). For example, learners would measure the length of a side of a triangle in degrees, employing Pythagoras' theorem to find the third angle of a triangle, and make connections in an idiosyncratic manner. In this case making 'connections' refers to linking a concept in a manner which is meaningful to the learner and conforms to its use in the mathematics community (Mhlolo & Schäfer, 2013).

Duval (1999) argues that when we fail to make connections with the representations produced, we succumb to what he refers to as a "blind spot of many didactical studies" (p. 20). For Duval, visualisation is the grasp of the whole. Duval (2013) posits visualisation as a cognitive activity and he proposes three kinds of cognitive activities in geometry: first, seeing and recognising shapes; second, measuring, calculating and comparing magnitude; and, third, inferring from properties. Many students, for example, fail to recognise a square or a rectangle when it is tilted. The process of visualisation could help students to realise that a square is not only an image, but also a shape controlled by its definition (or properties), irrespective of its orientation. Mathematical visualisation lies in the implicit selection of visual units within the configuration that are relevant, rather than ones that are not. Geometrical properties correspond to what remains invariant when the drawing is changed by moving either one of its points or its segments. Visualisation in geometry is specifically related to the cognitive activity of seeing shapes within a figure. Therefore, Duval (2013) argues that we can promote processes of visualisation among learners by making them identify the various line segments and points of intersection in any geometrical figure.

Visualisation and DGS

Many researchers regard ICT as a potentially powerful resource for learning (Ruthven et al., 2008; Stols & Kriek, 2011). When used appropriately, ICT can bring about an exciting paradigm shift in mathematics education, which allows for powerful multiple representations of mathematical concepts as visual objects. The South African National Curriculum Statement (Department of Basic Education, 2011) recognises the visualisation potential of ICT in education, and specifically recommends that learners should be able to use science and technology effectively and harness its visual and symbolic opportunities.

Arcavi (2003) recognises the potential role of technology in visualisation. He argues that technology as a visual aid creates an entry point for learning a mathematical concept. Stols and Kriek (2011) claim that "software has the potential to enrich visualisation of geometry" (p. 137). DGS provides tools which allow learners to engage with and control the inherent actions of the objects by pointing, clicking and dragging aspects of the shape on the computer screen. By dragging and moving components of any representation, the DGS provides visual validation of the properties of its invariant and variant elements. The tools of DGS, like dragging, enable learners to move shapes around the computer screen and thus discover for themselves their inherent characteristics.

The focus of this chapter is DGS in general and *GeoGebra* in particular. *GeoGebra* is an open-source and multiplatform dynamic mathematics software that affords a bidirectional combination of geometry and algebra. The bidirectional combination implies that by typing an equation in the algebra window, the graph (or geometric object like a circle, ellipse, etc.) of the equation will appear in the graphics window. Similarly, while dragging the graph (or the geometric object), the equation in the algebra window changes accordingly. The software runs on virtually any operating system, like *Windows*, *Macintosh*, *Linux* or *Android*, as it requires only a *Java* plug-in and, unlike *GeoGebra* commercial products, learners and teachers are not constrained by licences to run the software on a limited number of computers.

In this chapter, screenshots are used to convey the sense of the mathematical activities that the participants engaged in. I argue that the usual static and monochrome images do not always fully convey the essence that the dynamic and colour-coded applets used in the research do.

Ruthven et al. (2008) consider DGS a powerful medium for exploring and testing hypotheses. The dynamic visualisation with DGS helps learners to develop understanding, as they can construct, explore and test their own conjectures and hypotheses. An important feature of DGS is that it can be designed around mathematical principles and inspire students to interact with them. When the emphasis of a geometry lesson is on promoting students' broad understanding, then DGS can be used to develop awareness of mathematical ideas through exploring geometric figures dynamically. It must be emphasised, however, that technology-enabled visualisation is not believed to make mathematical learning easier, but rather as a basis for making mathematical learning richer and deeper.

Mathematical Proficiency in Relation to Visualisation

Kilpatrick et al. (2001) developed a comprehensive framework for mathematical learning proficiency. They identified a set of five intertwined strands that constitute mathematical proficiency. Due to the scope of this chapter, only one strand of mathematical proficiency is privileged here, namely, conceptual understanding.

According to Kilpatrick et al. (2001), conceptual understanding is the comprehension of mathematical concepts, operations and relations. It refers to the interrelated grasp of fundamental mathematical ideas, and refers to far more than merely the isolated facts and procedures. Conceptual understanding is conceived as relating new ideas to an existing knowledge structure. Thus, knowledge grows when isolated facts and propositions are linked and connected together to form an inter-connected network. Presmeg (1991, 2006) succinctly argues that the essence of visualisation lies in making connections. Understanding is characterised by a knowledge base that is connected to a rich supply of knowledge about concepts.

Furthermore, Arcavi (2003) sees visualisation as a process of constructing knowledge that promotes understanding. Duval (2013) asserts that as understanding involves grasping the whole structure, there is no understanding without visualisation. While a single activity focuses on one or some units and properties of a mathematical concept, visualisation leads to "grasping directly the whole configuration of relation and in discriminating what is relevant in it" (Duval, 1999, p. 13). The use of visualisation enables learners to understand mathematical ideas conceptually.

The Case Study

The research project that is at the heart of this chapter is a case study within the GLIP project, in which DGS formed an integral part of the pedagogical arrangement. This case study took place during regular Grade 11 mathematics classes. A cohort of six, mixed gender, Grade 11 learners from two schools in Mthatha, who agreed to participate through voluntary informed consent, constituted the research participants for this case study. The unit of analysis was the individual learner's interactions with DGS. The research mainly used a combination of screen-capturing videos and stimulated-recall interviews as its data collection techniques.

Screen-capturing is a relatively novel method of data gathering in the field of education. Screen-capturing software captures everything that a user does on a computer screen, including the use of media files and any other related digital activities. My pilot study guided me on the technical logistics of the hardware and software required for this technique. The participants were informed that their workings on the computer screen would be recorded by a screen-capturing process.

I used stimulated-recall interviews with the individual participants to review the individual screen-capturing videos, in order to facilitate an accurate and robust reflective process. Owing to the scope of this chapter, that part of the larger study which analysed other topics and applets that the learners interacted with is not reported on.

The screen-capturing videos of the lessons and the stimulated-recall interviews were transcribed. The learner's actions and remarks were analysed inductively, according to codes derived from Kilpatrick's et al. (2001) proficiency framework, which is a larger study of which this chapter forms a part; these will not be reported on here, due to space constraints.

The Lesson Design

In one of the GLIP sessions the *angle at the centre theorem* was discussed and the GLIP teachers collaboratively designed numerous applets on the conceptual development of different aspects of the theorem. All the applets are accessible at www.visual-maths.com.

In Figure 8.1, Applet 1 was used for the learners to identify the angles at the centre and the angle on the circumference, and then, by dragging point B along the circumference of the circle, explore the relationship between them. The points on the circumference of the circle could be dragged in either direction along the circumference of the circle, but the ratio of the two angles was invariant.

In Applet 2, two aspects of the theorem were highlighted. First, that the angle on the circumference should be on the same side of the arc; and, second, that the *angle at the centre theorem* was linked to the *opposite angles of a cyclic quadrilateral theorem.*

Applet 3 has multiple uses such as exploring the ratio between the angles at the centre and on the circumference, highlighting the *same-side-of-the-arc* concept and the *angle subtended by a diameter.* The third applet was specially designed so that the angle at the centre could be changed (using a slider tool) by unit degrees.

The teachers provided their learners with questions to solve the angles, specifically applying the *angle at the centre* theorem, but not limited to it. Two questions and their sub-questions were selected from a textbook.

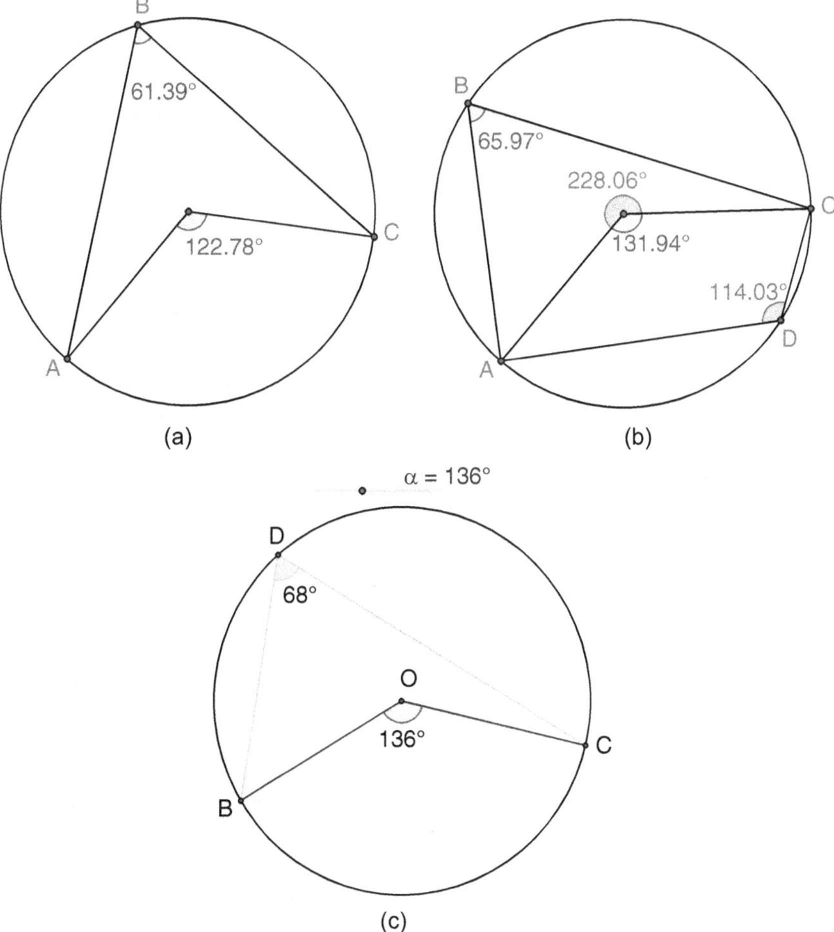

Figure 8.1 Three applets collaboratively designed by the GLIP teachers.

Unfolding of Lessons

The participant learners in this case study were Rose and Daisy from Class A, Jasmine and Aster from Class B and Cypress and Iris from Class C. The teacher in Class A allowed the learners to do constructions and reproduce Applet 1 and Applet 2, while the teacher in Class B chose to use his own pre-designed applets. The themes I discuss below emerged during my analysis, across all the participants – they are grounded in concrete incidences that were observed when the learners engaged with *GeoGebra*. These themes are pertinent to the meaning-making process of my participants. The themes identified are: (1) gaining knowledge beyond print and (2) internalising the theorem.

Theme 1: Gaining Knowledge beyond Print

The following five vignettes illustrate what happened in Classes A, B and C:

Vignette 1 – Constructions

The participating learners from classes A and C (Rose, Daisy, Iris and Cypress) constructed geometric objects in the DGS environment following their teacher's instructions. They initially struggled to construct circles and segments and measure the angles, but they acquired the skills swiftly as the class progressed. In the beginning, some of them had strange diagrams on their screens. Iris, Cypress and Daisy were not able to plot points exactly on the circumference of the circle. Instead of choosing the line segment tool, Cypress chose a line tool and had lines all over the screen. There were many technical mistakes made by the learners. Measuring angles posed a challenge to many learners. For example, selecting the appropriate points in the incorrect order was a major reason for not obtaining the required angles. For instance, Iris measured the reflex angles, and Cypress and Daisy measured the angles at the arms of the angle at the centre instead of at the vertex. However, in spite of this, as the lesson progressed, the participants succeeded in constructing useful and meaningful diagrams on their desktop computers.

Interestingly, in Class B, Aster independently explored the available tools in *GeoGebra* and started to draw and construct figures using Applet 3 (the teacher of class B allowed only the learners to use his pre-designed applets). Aster constructed circles and a triangle in one of the circles, as shown in the screen-capture in Figure 8.2. He later explained that he wanted to verify whether the angle subtended by a diameter was a right angle. Nonetheless, he could not proceed further with his investigation as he could not measure the lengths and angles using the *GeoGebra* tools.

A circle constructed in a typical *GeoGebra* worksheet involves defining three points on the circumference of the circle. This is not necessarily so for a circle drawn with a pencil and pair of compasses on a piece of paper. Another way of constructing a circle in *GeoGebra* involves identifying the centre point of the circle, and then drawing the locus with respect to this point. To measure the angles of a vertex at the circumference of a circle, a further set of points and lines must be made explicit in *GeoGebra*. Anything beyond a trivial construction involves a precise degree of accuracy with *GeoGebra*, which may not be necessary with pencil and paper. This analysis of learners' constructions of geometric figures in DGS shows striking similarities with the work of Ruthven (2003, p. 13), who recognises that DGS constructions "introduce[s] a degree of mathematical discipline into the use of a DGS, which is potentially educationally beneficial".

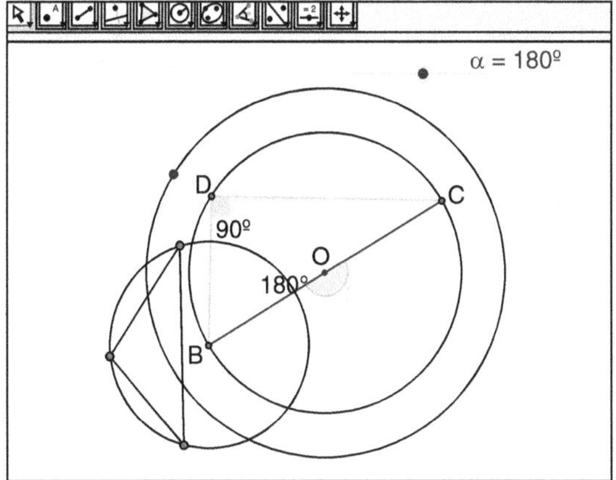

Figure 8.2 Aster constructed circles and triangles but could not measure the lengths of the lines and the angles.

Vignette 2 – Privileging dynamic visualisation

DGS, with its dynamic capabilities, provides an opportunity for learners to learn beyond printed textbooks. In Vignette 2, Rose (by playing and exploring with point B) arrives at a conclusion that the point B must be on the circumference in order for the *angle at the centre* theorem to hold. Rose investigated this by dragging the point E all around the screen, placing it inside the circle, at the centre, on the circumference and outside the circle (see the four screen-captures in Figure 8.3). Based on her engagement with the tools of *GeoGebra,* Rose concluded that the angle inside or outside the circle was not half the angle at the centre of the circle: only the angle that was formed on the circumference of the circle was half the angle at the centre of the circle. In her reflective account of the lesson, she observed that the textbooks talked about the points on the circumference as though they were static. Rose commented:

> Even if they are subtended by the same chord, the centre and circumference. If that is not at the circumference and around the circle or outside the circle then it will not be half the angle at the centre. I observed that the textbook did not really point this out.

Interestingly, Jasmine also observed something similar in her reflective interview after the lesson. She said that one cannot draw multiple figures on a chalk board or paper and then drag them to visualise or illustrate the theorem.

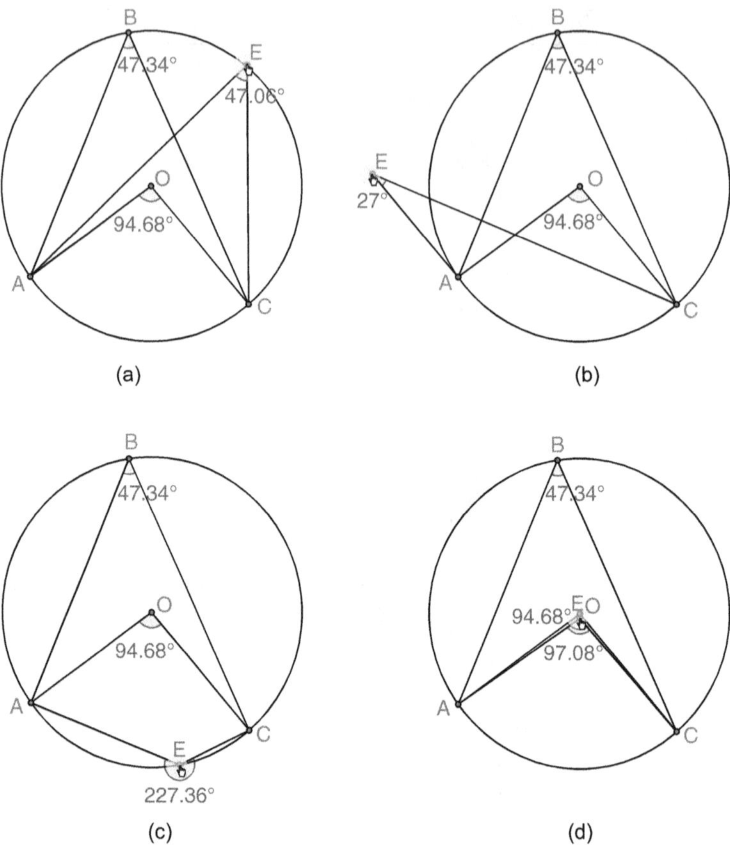

Figure 8.3 Rose dragged the point E around the screen.

> In chalkboard and paper you cannot move angles, but in computers you can. ... it's cool, it's really nice because it got me to understand things much better and put them into context than in the classroom.

The analysis shows that visualisation is a dynamic process that is difficult to replace with the written word, as "the written word, a static vehicle is not well adapted to the needs of the visualization processes" (De Guzman, 2002, p. 14).

Vignette 3 – Overwhelming rigidity

It was interesting to observe that all the participants, except Rose and Jasmine, consciously dragged points around, discovering and attempting to understand the concepts, before endeavouring to solve the given problems while interacting with one of the practice applets. For example, Daisy deleted one of the points on the circumference by mistake, while Iris dragged the points on the circumference of the circle and the figure became distorted. Cypress dragged the whole polygon inscribed in the circle. It was apparent that he was not able to solve the angles until the teacher guided him. However, Aster, Iris and Daisy settled down when they dragged the points on the circumference to match the standard diagram of the *angle at the centre* theorem, as shown in the sequence.

Once the participant learners conceptualised the idea, they could swiftly solve the sizes of the angles. Discussing the typological differences in mathematical abilities, Krutetskii (1976) describes the 'geometric type' as a learner who may start slowly but can progress quickly in learning a task and attaining a stable formation of ideas. In this case, the participants developed their understanding and reasoning slowly, step by step, "omitting no flight of thought" (Krutetskii, 1976, p. 192).

Bishop (1980) refers to 'geometric rigidity' as a trait in a learner who is unable to 'see' a diagram in a different way and from a different perspective. A good example is the standard diagram for the *angle at the centre* theorem where the angle at the centre of the circle is obtuse. Learners often find it difficult to 'see' that this angle is twice the subtended angle at the circumference. Nonetheless, the analyses of my lessons show that using dynamic geometry has the potential to obviate some of the difficulties associated with 'geometric rigidity'.

Vignette 4 – Real-life examples

During the reflective interviews, I specifically asked the participants to provide real-life examples of a circle and/or the *angle at the centre* theorem. All the participants listed similar examples of circles – buttons, wheels, bottle caps and pizzas. Interestingly, it was only Daisy who listed the computer as another source where she encountered the '*angle at the centre*' idea. Jasmine said:

> I know I will find it (moving her eyes all around the room)... in this school cameras are actually at the angle at the centre it is the point where it actually capture. Is the point where angle at the centre that captures. It can capture only within a particular radius....
>
> ... I can see the angle formed by the hands of hour hand and minutes hand. When I am moving the arc when teacher asked to do it, I think of seconds hand and minutes hand moving forwards and backwards.

The real-life examples provided by Jasmine were pertinent, as they demonstrated out-of-the-box, original thinking. She likened the circle to a movie set, where the centre of the circle represents the camera and the circumference of the circle, the movie scene. As the camera pans around the movie set (i.e. the circumference of the circle), the 'arc shot' describes the angular extent of the scene being captured. Also, there exists a circle passing through the location of the camera and two points on the movie set. Jasmine was also reminded of a clock, where the hour hand and minute hand formed angles at the centre of the clock face.

Vignette 5 – Proposition: Inversely proportional relation

It was intriguing that Iris came up with her own theory of points inside and outside the circle. She dragged the points to obtain different measurement for the angles, as shown in the two screen-captures in Figure 8.4. During her interaction with the teacher, using her own constructed figure in *GeoGebra*, Iris observed that there was no obvious relation between the angle outside and the angle 'hanging' inside. She posited a theory of her own – '*the farther the angle from the centre is, the smaller it becomes*'. The angle subtended by an arc of a circle becomes smaller in size as it moves away from the centre. Iris remarked:

> There is only one relation among the angles which is between angle at the centre and angle at the circumference. ... the angle at the centre is twice as big as the angle at the circumference. But the there is no clear relation

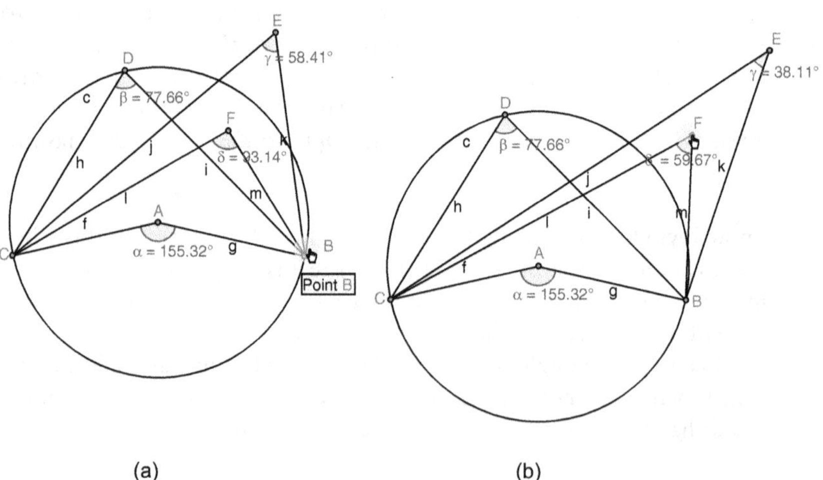

Figure 8.4 Iris dragged the points on the circumference.

> between angle inside the circle and angle hanging outside the circle. The higher you go the smaller it becomes.

This theory may not be relevant in the context of the Grade 11 school curriculum, but it is a legitimate mathematical observation and illustrates that the experience with the *GeoGebra* software enabled learners to visualise phenomena that their textbook illustrations did not achieve.

Theme 2: Internalising the Theorem

Jasmine felt that working with *GeoGebra* helped her to internalise particular theorems. She could understand how the theorems linked to one another. She created her own knowledge with little reliance on her teacher. Jasmine said:

> If I summarise, I would say it's informative. It was nice. It helped me to internalise most of those theorems. ... It also helped me to independently think and not always ask questions to the teacher.

The following three vignettes (6 to 8) illustrate Theme 2.

Vignette 6 – A particular case of the theorem

Aster's initial understanding was that the *angle subtended by a diameter* theorem did not need a proof. Given the visual stimulus of using DGS and the placing of previously learnt geometric concepts and theorems in context, Aster was able to explain how the *angle subtended by a diameter* theorem is a consequence of the other theorem. While interacting with a pre-designed applet, Aster dragged the points to make a cyclic quadrilateral into a triangle. The applet was intended to allow visualisation and generate the property of the sum of interior opposite angles of a cyclic quadrilateral. Instead of visualising a cyclic quadrilateral, Aster dragged a point over the others at various positions to form triangles, including a right-angled triangle, as shown in the two screen-captures in Figure 8.5. He established that other triangles formed on the circumference of the circle are right-angled, as shown in the exchange below:

ASTER: *Oh yeah yes now I remember. The angle formed by a diameter at the centre is 180°; therefore its half will be on the circumference; therefore, the angle subtended by a diameter on the circumference is 90° because it forms a straight line with the centre.*

RESEARCHER: *What else you see?*

ASTER: *When you look at the triangle ABC, (right angled at C) AB is the hypotenuse. But when you look at the circle AB is the diameter. They are same.*

RESEARCHER: *Can you explain?*

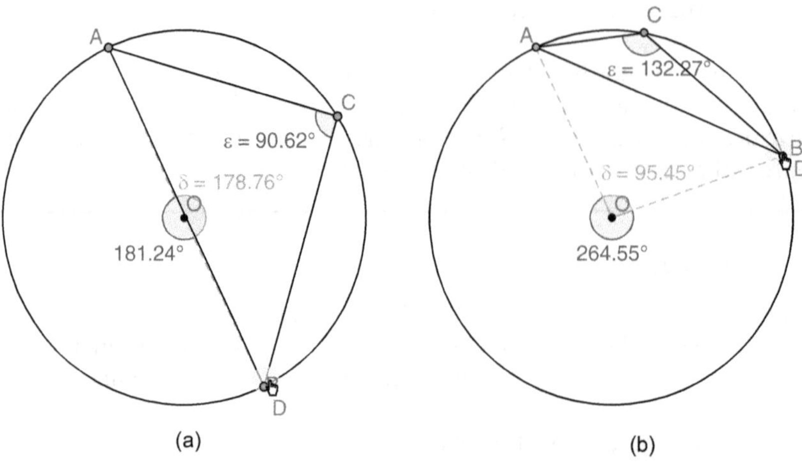

Figure 8.5 Aster dragged points to make triangles from the cyclic quadrilateral.

ASTER: *I can draw a circle around the right triangle and hypotenuse will be the diameter.*

An analysis of Aster's understanding of theorems raises two perspectives. First, Aster's unformulated exploration of theorems using the applet ultimately led him to unpack the hidden mathematical structures of the *angle at the centre* theorem. Aster visualised the *angle subtended in a semi-circle* theorem as a special case of the *angle at the centre* theorem. Second, despite Aster's initial scepticism of the connection among theorems, it seems that it was precisely his active engagement with the applets (as seen in the screen-captures in Figure 8.5) that led him to the development of the 'hypotenuse-diameter equivalence'. He visualised the angle subtended by a diameter as a right-angled triangle – a direct relationship of the diameter with the hypotenuse. However, he experimented further and found that the reverse relationship of the hypotenuse with the diameter was also true. Aster arrived at the conclusion through an explorative interaction with the pre-designed applets.

Vignette 7 – Conjecturing

Daisy discovered the *same-side-of-the arc* concept through construction and measurement. The four screen-captures in Figure 8.6 indicate the sequence of her measuring the angle at the centre related to $C\hat{F}A$. After measuring the inside angle at point F, Daisy was asked by the teacher to identify and measure the angle that would be double $C\hat{F}A$. But Daisy started incorrectly

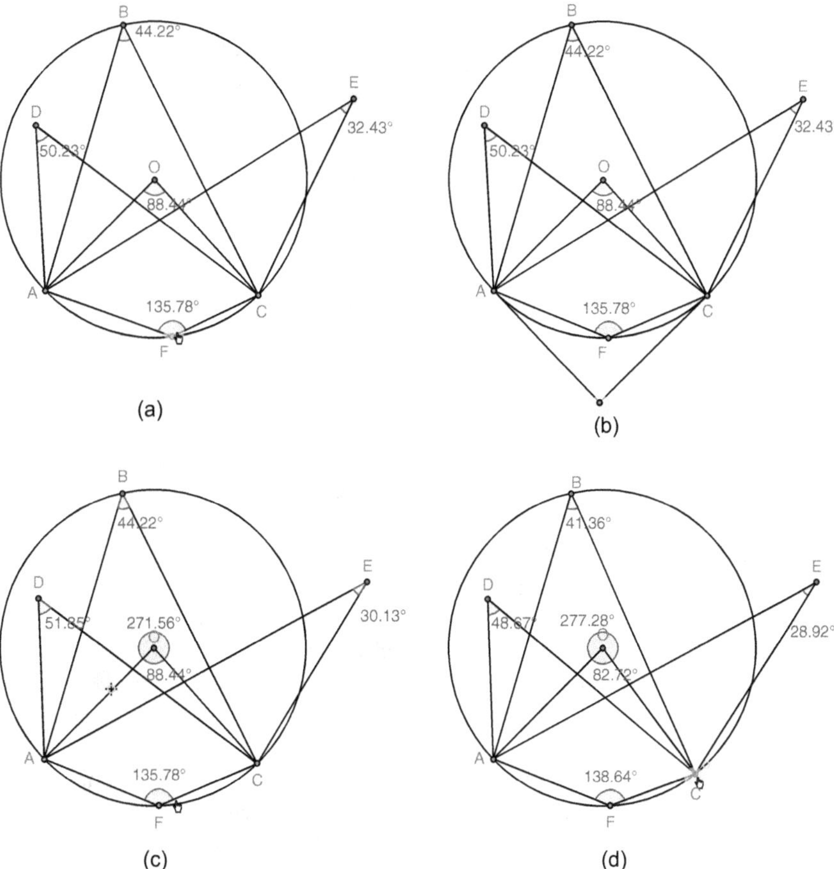

Figure 8.6 Daisy measured the angle at the centre related to $\hat{F}$.

by constructing an angle outside the circle. Daisy had a different conception of the angle at the centre; hence she drew a point outside the circle. Her construction disproved her conjecture and she realised that the angle should be inside the circle and at the centre of the same circle. She then measured the angle at the centre and appropriately related it to the angle $C\hat{F}A$. She then dragged the point F around the minor arc AC, continuing to confirm whether it was still half the angle at the centre. As she was engrossed in constructing her figures, reflecting and experimenting with her conjectures, she was able to resolve her uncertainty. Her manipulations with the geometric objects led her to visualise and conceptualise the appropriate angle at the centre.

Vignette 8 – Proving the theorem

In Euclidean geometry, we find rich connections among and between concepts and theorems. Duval (1999, p. 13) supports this by saying that "visualization (in mathematics) consists of grasping directly the whole configuration of relations and in discriminating what is relevant in it". The participants in my study sought to identify relevant geometric concepts. They learnt to recognise variant and invariant properties in varying degrees.

It is intriguing to analyse the observations of Cypress, Aster and Daisy. They recognised that the *angle formed at the centre by a chord resulted in an isosceles triangle, owing to equal radii*. However, Daisy's strategy was not broad enough when compared with the strategies of Cypress and Aster. Daisy saw only one isosceles triangle ΔAOE, whilst Cypress and Aster saw two isosceles triangles, including ΔBOE. They were further able to make astute deductions that $\overline{BOA}$ made an exterior angle with the ΔBOE, and applied two theorems of the triangle, *angles opposite to equal sides* and *exterior angle property of a triangle*, to arrive at the angle $A\hat{O}E$ – the angle at the centre subtended by the chord $\overline{AE}$.

During further investigations in the reflective interview, I solicited curiosity by asking: "*Why is the angle at the centre double that of angle at the circumference, and not treble or five times?*" Daisy referred to her experience and observation while dragging the points on the circle: "*I think the centre part is a revolution and the circumference you can change the points and we noticed that it is two times. But I am not sure*".

Daisy was able to produce numerous angles using *GeoGebra*, and dynamically verified the theorem. There was little need for her further conviction.

However, Cypress and Aster contemplated that the radius of the circle made the *angle at the centre* theorem. Apparently, they were referring to the construction of points and angles inside and outside the circle. Their contribution is valuable in the sense that constructing counter-examples is useful for verifying conjectures. Cypress observed: "*When the point is not on the centre, there is no radius, therefore it cannot be double*".

Nonetheless, after sufficient scaffolding, Aster managed to extend his visualisation capabilities to incorporate different concepts and theorems and managed to arrive at the theorem. He noted: "*The exterior angle of an isosceles triangle is twice the interior opposite angle*".

The participants appeared to prove theorems through investigations, attaining what De Villiers (1996, p. 28) describes as a "satisfactory sense of *illumination*, i.e. an insight or understanding into how it is a consequence of other familiar results". Through extensive constructions of circles, lines and angles, and the dragging of points, the learners learnt and discovered the results of the theorem to be true. However, it becomes necessary from a teacher's perspective to explain how this is a logical consequence of previously proven theorems.

Further Discussion

The main purpose of this chapter was to provide a brief analysis of the participating learners' interactions with *GeoGebra,* a DGS, on the topic of *angle at the centre* theorem. The results of the analysis indicate that there were similarities and differences across the participants in the development of their conceptual understanding of the theorem, and, as the analysis unfolded, different characteristics of learning proficiency in relation to visualisation were evident.

Furthermore, the research highlighted how the participants engaged enthusiastically and critically in exploring and manipulating the given applets. In circle geometry, *GeoGebra,* with its dynamic possibilities, enabled the learners to focus their attention on continuous variations of angles as they dragged the vertices of angles to different positions on the computer screen. The participants were engrossed in their own constructions, reflecting and experimenting with their conjectures. More importantly, their manipulations with the geometric objects ultimately led them to conceptualise the relevant circle geometry theorems.

Role of Visualisation

Considering Duval's (2013) observation that the construction of geometrical figures by learners is crucial in visualisation, the use of DGS here proved to be an efficient visualisation tool to enable the learners' understanding. At the outset, in using the software, a solid understanding of geometric properties is essential, however. The use of software tools reinforces related geometric principles. The practical experience of geometric constructions in a DGS environment provides a context in which the conceptual development of geometry theorems may be achieved.

With its dynamic possibilities, *GeoGebra* enabled the learners to focus their attention on continuous variations of angles as they dragged the vertices of angles to different positions on the computer screen. In a pen-and-paper environment, constructing multiple static figures would be inefficient, but the dynamic visualisation capabilities of DGS enabled the participant learners to move from concrete instances to conjecture abstract ideas, echoing De Guzmán (2002), who observed that visualisation is an important aspect in mathematical activity, through which one can explore different structures of concrete reality. The dragging of the points highlighted the variant and invariant properties of the angles, supported by visual recognition of the relationships between them.

The dynamic nature of *GeoGebra* offered many opportunities for participants to make conjectures by analysing different configurations. The learners had an abundance of opportunities for exploring, conjecturing, refuting, reformulating and explaining. They deepened their understanding of the mathematical relationships involved, and sharpened their ability to validate their conjectures. De Villiers (1996) comments:

> Dynamic geometry software strongly encourages this kind of thinking as they are not only powerful means of verifying true conjectures, but also extremely valuable in constructing counter-examples for false conjectures. (p. 14)

For Presmeg (2014), visualisation is employed in classrooms when we make connections between topics in mathematics. The participant learners were able to see and connect theorems and not merely regard them as isolated facts and statements. For instance, *GeoGebra* with its drag mode capability enabled them to visualise the ***angle subtended by a diameter*** as a special case of the ***angle at the centre*** theorem and not as a discrete theorem. Working with this software, learners could relate how the theorems were linked to one another. I agree with other researchers (Arcavi, 2003; Ruthven et al., 2008) who accentuate that the visual trajectory in mathematics education is linked to the use of dynamic potentialities in computers.

Furthermore, the role of a teacher should be to effectively utilise opportunities for the explanation and investigation of geometric theorems, rather than using DGS only for verification and proof in geometry. Perhaps GLIP meetings in future should include discussions on how to explore and investigate geometry with or without using technological tools, with a special emphasis on making real-life connections.

Furthermore, when participants dragged the points, their attention focused on the continuous dynamic variation of figures and angles. This positioning of gaining ***knowledge beyond print*** is particularly interesting because it demonstrated different classroom opportunities "which tacitly bridged between dynamic and static versions of the 'same' mathematical result" (Ruthven et al., 2008, p. 314).

I contend that dynamic visualisation capabilities in *GeoGebra* help learners to develop understanding, as the visual feedback of their actions on applets develops new understanding. Technology-enabled visualisation, thus, is not regarded as making learning easier, but as a basis for enriching a mathematically activity.

Conclusion

The integration of a DGS environment, particularly the use of the visualisation potential of *GeoGebra* in mathematics classrooms, was the research focus of this chapter. This research is particularly significant as it made use of an empirical field and context that has been a victim of underdevelopment, inequality and woefully under-utilised educational resources.

With this case study, I wish to offer a contribution as to how DGS can be used meaningfully to mediate a visual approach to learning geometry. The results of the case study point to exciting opportunities for learners to develop dynamic visual reasoning abilities. In addition, a rewarding aspect of this research has been the use of screen-capturing software that recorded learners' activities on the computer. These videos provided me with the precise and accurate information required for my research. This method enabled me to gain insights into the participants' mathematical thinking and visualisation processes, and thus addressed my research objectives. Screen-capturing, as a research instrument to deconstruct learning habits and thoughts, is, in my experience, very powerful. From a research methodology perspective, I recommend that studies addressing the integration of technology into classrooms make use of screen-capturing software.

This research study was designed around the GLIP intervention programme, which meant that the teachers implemented what they learnt in the intervention directly in their classrooms. This was a very effective strategy and ensured that my research project remained grounded and had an impact on my immediate community. I thus recommend that the integration of technology in other mathematics classrooms be accompanied by a parallel intervention programme to ensure that on the basis of the teachers' and learners' reflections, appropriate and immediate support can be actioned to ensure a smooth and coherent integration process.

References

Arcavi, A. (2003). The role of visual representations in the learning of mathematics. *Educational Studies in Mathematics, 52*(3), 215–241.

Bishop, A. J. (1980). Spatial abilities and mathematics education : A review. *Educational Studies in Mathematics, 11*(3), 257–269.

De Guzmán, M. (2002). The role of visualization in the teaching and learning of mathematical analysis. In *Proceedings of the International Conference on the Teaching of Mathematics* (Vol. 7, pp. 1–25). Crete, Greece. https://doi.org/10.1109/IEMBS.1992.5761740

De Villiers, M. (1996). The future of secondary school geometry. In *SOSI Geometry Imperfect Conference* (pp. 1–35). 2–4 October, UNISA, Pretoria.

Department of Basic Education. (2011). *Curriculum and Assessment Policy Statement (CAPS)*. South Africa. Government printer.

Department of Basic Education Ministerial Committee. (2013). *Investigation Into the Implementation Of Maths, Science And Technology*. South Africa.

Duval, R. (1999). Representation, vision and visualization: Cognitive functions in mathematical thinking. Basic issues for learning. In *Twenty First Annual Meeting of the North American Chapter of the International Group for the Psychology of Mathematics Education* (Vol. 25, pp. 3–26). https://doi.org/10.1076/noph.25.1.3.7140

Duval, R. (2013). The first crucial point in geometry learning: Visualization. *Mediterranean Journal for Research in Mathematics Education, 12*(1, 2), 23–37.

Hilbert, D. (1952). *Geometry and the Imagination*. (Translated by P. Nemenyi, Ed.), (1990 ed.). Chelsea Publishing Company.

Kilpatrick, J., Swafford, J., & Findell, B. (2001). *Adding it Up: Helping Children Learn Mathematics*. Washington: National Academy Press.

Krutetskii, V. A. (1976). *The Psychology of Mathematical Abilities in Schoolchidren*. (Transalated from the Russian by Joan Teller, Ed.). Chicago: Univeristy of Chicago.

Mavani, D., Mavani, B., & Schäfer, M. (2018). A case study of two selected teachers as they integrated dynamic geometry software as a visualisation tool in teaching geometry.

African Journal of Research in Mathematics, Science and Technology Education, *0*(0), 1–11. https://doi.org/10.1080/18117295.2018.1522716

Mhlolo, M. K., & Schäfer, M. (2013). ' The area of a triangle is 180° C '— An analysis of learners ' idiosyncratic geometry responses through the lenses of Vygotsky' s Theory of Concept Formation. *African Journal of Research in Mathematics, Science and Technology Education*, *17*(1, 2), 83–93. https://doi.org/10.1080/10288457.2013.826973

Mudaly, V. (2010). Thinking with diagrams whilst writing with words. *Pythagoras*, *75*(July), 65–75.

Presmeg, N. (1986). Visualisation in high school mathematics. *For the Learning of Mathematics*, *6*(3), 42–46.

Presmeg, N. (1991). Classroom aspects which influence use of visual imagery in high school mathematics. In Furinghetti, F. (Ed.), *Proceedings of the 15th PME International Conference* (pp. 191–198). Assisi.

Presmeg, N. (2014). Contemplating visualization as an epistemological learning tool in mathematics. *ZDM - International Journal on Mathematics Education*, *46*(1), 151–157.

Ruthven, K. (2003). Linking algebraic and geometric reasoning with dynamic geometry software. *Final Report to the Qualifications and Curriculum Authority*. (March) Cambridge University.

Ruthven, K., Hennessy, S., & Deaney, R. (2008). Constructions of dynamic geometry: A study of the interpretative flexibility of educational software in classroom practice. *Computers and Education*, *51*(1), 297–317. https://doi.org/10.1016/j.compedu.2007.05.013

Stols, G., & Kriek, J. (2011). Why don't all maths teachers use dynamic geometry software in their classrooms? *Australasian Journal of Educational Technology*, *27*(1), 137–151.

Stols, G., Mji, A., & Wessels, D. (2008). The potential of teacher development with Geometer's Sketchpad. *Pythagoras*, 68, 15–21. https://doi.org/10.4102/pythagoras.v0i68.63

9 Teaching and Learning with Mobile Technologies

Clemence Chikiwa and Matthias Ludwig

Introduction

One of the most important goals of mathematics at all levels of learning is to understand the thinking involved in teaching, learning and doing mathematics, in an effort to enhance and improve access to knowledge. Studies have shown that the learning and practice of mathematics are not purely and solely intellectual activities isolated from social, cultural and contextual factors (Cahyono, 2018; Daniel, 2020; Vygotsky, 1978). These activities are part of communities in which teachers and learners live. Mathematics is regarded as a human activity that has a great deal to do with where and how people live. Consequently, a major component of children's mathematical development is the ability to apply their learning to real-world contexts (Ludwig & Jablonski, 2019; Nardi, 2014).

One approach to blend teaching and learning with the real world is by using outdoor activities. Teaching and learning mathematics in outdoor environments provides endless opportunities for increasing epistemological access and enhancing conceptual understanding of mathematics concepts. Many mathematics-rich examples from both living and non-living objects can be found around us. The home environment, school grounds, shopping mall, neighbourhood park, local garden centre and city museum, among many others, are all brilliant examples of spaces for learning mathematics. The use of such spaces provides a visual representation of mathematical concepts that in some cases are viewed as abstract and detached from real-life experiences. Mathematics need not be restricted to the classroom. Nobody should assume that education can only be carried out in a rectangular room with our eyes focused on a chalkboard. This chapter aims to explore the use of outdoor activities and smartphones as visualisation and mathematising tools by teachers and learners, through mathematics trails developed using the MathCityMap (MCM) project in South Africa, Namibia and Germany.

Outdoor Activities and Visualisation

Outdoor teaching and learning of mathematics is an important component in mathematics education (Cahyono, 2018), due to various reasons that are tied to educational, historical, political, economic and social settings. The coronavirus

DOI: 10.4324/9781003172420-12

(COVID-19) that was declared a worldwide pandemic has also led to the reconsideration of how teaching and learning can be approached in ways that minimise contact between humans (Barlovits & Ludwig, 2021). We argue that the teaching and learning of mathematics can be enhanced and enriched by incorporating outdoor activities in conjunction with technological developments that are now increasingly available to both teachers and learners. In this chapter we focus on the teaching and learning of mathematics using smartphones and outdoor activities as visualisation tools.

The use of locally available objects in the learners' immediate environments may provide advantages that are associated with visualisation in teaching and learning. Arcavi (2003) defines visualisation as:

> the ability, the process and the product of creation, interpretation, use of and reflection upon pictures, images, diagrams, in our minds, on paper, or with technological tools, with the purpose of depicting and communicating information, thinking about and developing previously unknown ideas and advancing understandings.
>
> (p. 217)

Visualisation is thus a process that uses images, diagrams and actual artefacts, with or without technology, to advance mathematical understanding. According to Arcavi (2003) and Nardi (2014), the process of visualisation can be external or internal. External features include the use of visual artefacts such as physical manipulatives, drawings, charts and diagrams. Actual objects, both living and non-living, such as buildings, trees, roads, pavements and playing fields, among others are also included. Other external features are dynamic representations such as animations, for example, virtual manipulatives. Gutiérrez (1996) posits that some mental images are created from sensory cognition of concrete materials that learners engage with when using their sense organs. These images may then be expressed or communicated in the form of diagrams, pictures, drawings, gestures and discussions. This may result in a visual approach to teaching and learning mathematics that has the potential to make obvious how and why mathematics works (Makina, 2010).

The internal process of visualisation involves the formation and manipulation of mental images (Cohen & Hegarty, 2007). As pointed out by Makina (2010), visualisation incorporates those internal processes that are rooted in concrete experiences in the mind. These processes make use of, or are characterised by, visual imagery, visual memory, visual processing, visual relationships, visual attention and visual imagination. Arcavi (2003) further said that "the visual display of information enables us to 'see' the story, to envision some cause-effect relationships, and possibly to remember it vividly" (p. 218). We thus argue that when teaching and learning employs actual objects in the environment or images on paper or technological devices, learners' knowledge of concepts becomes grounded and is retained for a longer period than when only verbal communication is used during teaching and learning (Makina, 2010). Both external and internal visualisation processes are important and relate to each other. This is because, according to Vygotsky (1978), learning starts on an external plane before it is internalised – that is, incorporated into the internal plane. Outdoor activities are thus considered in this study as providing appropriate visual external features that will eventually lead to the internalisation of mathematical concepts. These outdoor activities, as explained later in the chapter, are accessed through the use of smartphones that have an inbuilt GPS (Global Positioning System) system.

Use of Smartphones in Teaching and Learning

Over the years, the smartphone has become a multipurpose mobile device in many societies. A mobile phone is considered a handheld computer with an intrinsic connection to mobile networks (Sharples et al., 2010). Its ability to access mobile networks gives it many advantages as a visual tool that can be used for teaching and learning mathematics in and outside the classroom. The development of smartphones and their increased availability have had significant influence on the education system. Smartphones are able to provide suitable learning platforms through applications that learners may use in environments other than the classroom for their academic activities (Fabian et al., 2018).

Online teaching and learning have increased in the last few years due to various reasons. ICT devices, particularly smartphones, have become ubiquitous in most educational environments. However, some schools in Southern Africa see this as problematic. Many schools have even initiated policies that ban smartphones on school grounds. This has led to schools experiencing challenges of whether or not to permit the use of smartphones during teaching and learning. Such policies are now being challenged by the COVID-19 pandemic environment which has forced schools to adapt their teaching and learning processes to include smartphone technology.

General use of smartphones, especially by the younger generation, has been expanding across sub-Saharan Africa over the years in both urban and rural contexts (Porter et al., 2012). Such growth in use has led to some countries shifting from policies that totally ban the use of smartphones in schools by learners to policies that promote controlled use of these devices. South Africa is one such country where the use of smartphones in schools by learners is not banned at a national level, but schools are encouraged to monitor and control their use. It is also noted that in 2019 South Africa had a population of 58.56 million with 101.9 million mobile subscribers across five main mobile network operators – resulting in a unique subscriber penetration rate of 170%[1] (McCrocklin, 2021), implying that there were more mobile subscriptions than the whole population. Until recently, in other Southern African countries such as Namibia, the use of smartphones in schools by learners was prohibited. This was despite a mobile telephony penetration of over 110% (Statista, 2021). In Germany, on the other hand, there is no general ban on smartphones in schools for learning purposes, but they are regulated and "the smartphone phone penetration among youth up to 19 years of age is close to 100%" (MPFS, JIM-Study, 2020, p. 8).

While policies in Southern African education permit controlled use of smartphones for teaching and learning, these mobile devices are still viewed as an antithesis to traditional education and their use is banned in the majority of schools (EduTech Tours, 2017). It is, however, left up to individual schools to decide whether to allow or not allow the use of mobile devices during teaching and learning. In Namibia, the policy guidelines for Namibian education prohibits learners from bringing smartphones and tablets into classrooms (Namibia Press Agency, 2007). Learners' use of these devices on school premises is still restricted rather than regulated, despite findings from research, such as the exploratory research study conducted in five regions in Namibia by UNICEF (UNICEF, 2016) reporting an increased use of smartphones and the internet by children under the age of 16.

It is argued that the use of smartphones in schools for educational purposes can be advantageous in a number of ways. Mathematics tasks can be located anywhere within the learners' environment and smartphones can be used to access such tasks at any time convenient to the learners and the teacher. Fabian et al. (2018) argue that smartphones are cost-effective and require less infrastructure than the commonly accepted desktop computers. It is thus argued that the smartphone has the potential to unlock the greatest potential for delivering ICT-based learning to marginalised learners (UNICEF, 2016). This, however, seems to contradict other writers (Duncan-Williams, 2020; Ngesi et al., 2018), who have argued that the use of mobile devices could be costly for most schools, especially in the poor rural areas of Southern Africa.

On the other hand, Ludwig and Jesberg (2015) assert that smartphones can be used in educational settings to create polished and worthwhile products such as learning tasks that can be shared, solved in groups, published and used to generate discussions. The ability of smartphones to allow mobility during learning (Zender & Ludwig, 2019) has advantages. First, it can promote visualisations rooted in evoked images of real everyday objects outside the classroom that learners have interacted with before. Second, it allows teaching and learning to be mobile and in various environs. This in turn may prompt learners to think about their own experiences and share with their peers while solving problems on outdoor trails. Smartphones provide learners with

> … the opportunity to construct their own mathematical knowledge by solving the prepared tasks on [a] math trail [for example], supported by their interaction with the environment, including the digital environment.
>
> (Cahyono, 2018, p. 43)

The other advantage of mobile phone use is their ability to encourage active learner interaction and integration with various environments. This allows learners to leave the boundaries of the classroom and engage with hands-on mathematics activities. The mobility afforded by smartphones can facilitate an active network of learning where learners move in and out of different learning spaces, investigating mathematics properties within their environment (Fabian et al.,

2018). This interaction with smartphones in a mathematics learning environment is made possible by mobile applications such as the MCM as discussed below.

The continuous rise in the use of devices such as smartphones and tablets has sparked a mobile learning revolution (Kukulska-Hulme et al., 2017) which seems to be gaining significant recognition in previously traditional schools. Research has since shown that there is ample evidence that mobile platforms can support quality teaching and learning experiences and have many benefits in educational settings (Duncan-Williams, 2020). Such a rise in the use of smartphones (even in poor rural and township schools and communities) needs to be taken advantage of in teaching and learning. This chapter thus argues for the use of mobile devices to promote outdoor mathematics teaching and learning in all communities. We are not naïve regarding the challenges associated with the incorporation of smartphones in schools. These include additional control and monitoring measures, anxiety, addiction, reduced face-to face quality communication and divided attention from learners, among others. We argue in this chapter that smartphones are already in increased use in the daily lives of learners and their teachers, and that this trend will continue into the unforeseeable future. Also, the current social measures associated with the COVID-19 pandemic – such as social distancing – makes the mobile phone an appropriate teaching and learning device. It is against this background that applications such as the MCM (www.mathcitymap.eu) are becoming useful for teaching and learning purposes. The next section looks at the MCM app and how it is used to encourage outdoor mathematics teaching and learning.

The MCM App for Teaching and Learning

The MCM project is one of the more recent, unique digital apps that can be used as a medium of instruction in a mobile learning environment when teaching mathematics. The MCM project is a two-component system. First, there is a web portal with a data base where users, mainly teachers, can create mathematics tasks and transform them into trails; and, second, there is a free app which interacts with the web portal which is used by the students. This app allows mathematics to be experienced outdoors through the use of trails composed of tasks that learners can follow. The award-winning project was developed in 2012 by the MATIS I Team from the Goethe University, Frankfurt, Germany (Cahyono et al., 2015). The MCM project is a protocol-based application centred on the principles of mathematics trails. Users of the app follow a set mathematics trails aided by a smartphone, stopping at strategically placed waypoints to solve a given mathematical task. The app provides hints and gives appropriate feedback to the users' solutions (see Figure 9.1). The MCM project uses the GPS to locate the trails, waypoints and tasks.

The MCM project brings into mathematics education innovative ways and ideas of combining mathematical outdoor tasks, through trails, with mobile technology (Ludwig & Jesberg, 2015). As noted by Cahyono et al. (2015), "the

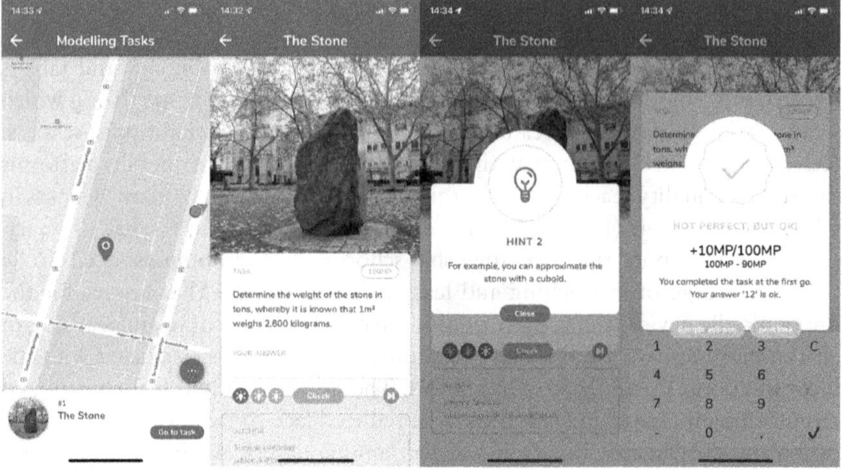

Figure 9.1 Map, task, hint and answer validation in the MCM app.

use of digital technologies in mobile learning environments has the potential to support teachers in facilitating outdoor mathematics teaching and learning processes" (p. 7). The mathematics that is learnt in the classroom (from textbooks and the chalkboard) is linked to reality through outdoor activities within real situations in the outdoor environment, through math trails.

Math Trails

A mathematics trail is a planned walking trail or path that contains a set number of outdoor tasks that are in the form of stops or stations (McDonald & Watson, 2010). Each stop on the trail contains at least one question that is to be attempted. Mathematics trails have the potential to arouse interest, by stimulating spontaneous observation among learners of the connections between mathematics and various geometric shapes and patterns found in the outside world. In this chapter, we argue that mathematics trails orient themselves within the constructivist paradigm because they encourage learners to discriminate, reason, communicate and solve problems, while simultaneously exploring, discovering and celebrating the beauty of mathematics and its presence. Cohen (2019) argues that since mathematics trails focus on the process of formulating, analysing, predicting and explaining, learners absorb and apply mathematical concepts naturally.

The idea of math trails originated some decades ago. In the 1980s, the first documented math trails were created in Melbourne by the Australian mathematics educators Blane and Clarke who prepared thought-provoking mathematics tasks that could be solved by families while on holiday (Blane & Clark, 1984). Thereafter, mathematics educators took advantage of the existence of these trails by including them in their instructional programmes. The MCM project made use of the already existing idea of mathematics trails in the educational context, with the help of new technologies (Ludwig & Jablonski, 2019). This project is worldwide and freely accessible. For downloading the intended and planned trails, internet access is required, but for the running of the trail at the preferred task location, no internet access is necessary.

The project has four broad components that users can take advantage of. These include outdoor tasks, mathematical city or outdoor trips, the map-based app and the MCM community. As explained by Ludwig and Jesberg (2015), the physical math trail tasks created by the task authors can be accessed via a mobile app bridged through a web portal. Summarised by Cahyono (2018), this portal is useful for four reasons: first, for receiving mathematical tasks; second, for storing created tasks so that they can be reused whenever the need arises; third, to create and store the routes that have locations with tasks; and, last, to provide at most three stepped hints for each task that authors have uploaded

These tasks are accessed through the GPS coordinates on a mathematics trail map. Trail data and task spots are available on the MCM app map. Moreover, the app gives direct feedback on entered solutions (Gurjanow et al., 2017), and offers opportunities for the user to ask for stepped hints (Ludwig & Jablonski, 2019) (see also Figure 9.1). These stepped hints are important during the learning process as they help learners if they fail at their first attempt to find the solution. During the solving of a given problem on site, the app provides learners with up to three opportunities to get a hint. According to Friedrich and Mandl (1992), the fulfilling purposes of stepped hints is motivation and enhancing visualisation. These stepped hints are a crucial component of the MCM project, and they are carefully developed to enhance learning. During the solving of an outdoor task, when learners acquire more opportunities to attempt a task, they are encouraged to employ various methods at their disposal. These include sketching, probing questions or engaging through physical contact with the task material, so that they seek more ways of representing the task problem, before going on to other tasks in the trail (Gurjanow et al., 2017). Research has shown that the exploration of outdoor tasks can become an adventurous, interesting and easy way of teaching, learning and experiencing mathematics (Fabian et al., 2018). If properly implemented, this can "stimulate student interest and engagement and the development of a healthy, accurate view of mathematics as a useful discipline" (Trafton et al., 2001, p. 264). Cahyono (2018) writes that "taking mathematics outside the classroom also allows learners to experience mathematization processes in solving problems around them and not only in textbooks" (p. 51). A mathematics trail is one of the approaches to mathematics that takes teaching and learning out of the classroom context.

One use of math trails is that the solving of MCM tasks by learners enables them to indirectly complete Blum and Leiss' (2007) seven-step modelling cycle (see Figure 9.2). For the purpose of this chapter, we understand modelling as the ability to construct mathematical problems and relevant questions from the environment (Blum et al., 2007), to mathematise, to solve outdoor tasks and,

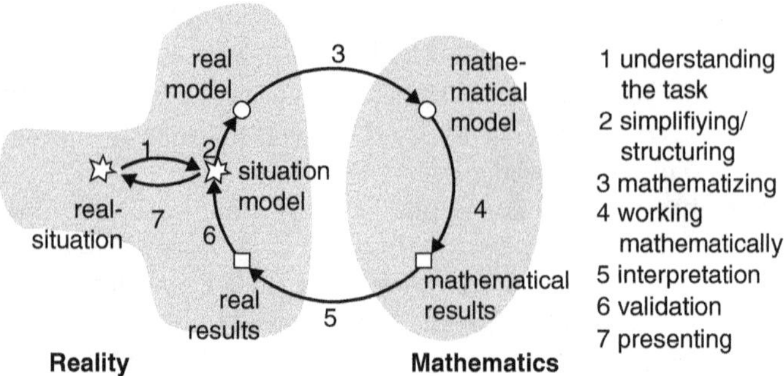

Figure 9.2 Blum & Leiss' (2007) modelling cycle.

finally, to validate and interpret the proposed solutions in the given real situation. Furthermore, the learners' ability to choose from different models and to evaluate them is also part of modelling ability (Blum et al., 2007). In particular, MCM tasks lay emphasis on the feature that allows different models to be chosen to address real-world problems.

In this chapter, we advance the idea that MCM trails are mainly questions that are particularly suitable for introducing modelling and should therefore be distinguished from complex and extensive modelling tasks.

The use of outdoor tasks in teaching and learning makes the exercise inherently visual in most cases. Mathematics concepts can be grounded in the everyday life of the teacher and the learner. When this is achieved, this results in making mathematics real and authentic to the learner. Currently, the general feeling that learners have towards mathematics is that it is an abstract subject that has very little to do with their everyday lives (Gafoor & Kurukkan, 2015; Langoban, 2020). This chapter argues that such a negative disposition can effectively be addressed when outdoor tasks are used to teach and learn some mathematics.

Theoretical Framing

This chapter is informed by *constructivist theory*, *embodied cognition* and *realistic mathematics education* (RME).

The proponents of mobile learning put forward that the activities used in mobile learning follow a *social constructivists'* ideology, especially when learners work in collaborative groups (Isenberg et al., 2011). This in turn promotes interactive learning which encourages and facilitates flexibility in the teaching and learning process and its relationship with the real world (Hayes & Scassellati, 2013). In this chapter, we take a theoretical approach stemming from the constructivist perspective, by positing that teachers can ground learning by engaging their learners in outdoor activities. Such activities, if well designed, can enable learners to connect their prior experiences, intuitions and understandings to real-life mathematical artefacts, representations and other tools (Piaget,

1952), to enable them to construct mathematical meanings upon their own personal understandings (Mariotti, 2009).

The second theoretical framework for this chapter is *embodied cognition* (Lakoff & Núñez, 2000). This theory has been chosen because it states that the comprehension of abstract mathematical concepts is rooted in sensory-motor experiences and in interaction with the environment and the world. We posit in this chapter that mathematics learning and cognition are and must be situated and context-dependent, through the use of outdoor activities created by the teacher and accessed via the MCM app. Research has since

> ... acknowledged that learning and teaching take place, and have always taken place, within embedding social contexts that do not just influence, but essentially determine the kinds of knowledge and practices that are constructed.
>
> (Núñez et al., 1999, p. 45)

According to embodied cognition theory (Shapiro, 2014), interaction with the physical world during learning influences even determines our cognitions (Kemmerer et al., 2013). In the realm of embodied cognition, sensory and motor systems are regarded as fundamentally integrated with cognitive processing. Thus, an embodied cognition approach attempts to provide continuity between the environment in which learners live and cognition.

Three of the principles of embodied cognition as listed by Wilson (2002) are important for this chapter. First is the notion that cognition is situated. Cognitive activity takes place in the context of a real-world environment in which the concerned individual lives, and it inherently involves perception and action. The second is that we offload cognitive work onto the environment. The human mind has limited information-processing abilities (e.g., limits on attention and working memory) so humans use the environment to reduce the cognitive workload. This makes learning and cognition situated as humans use the environment to hold or even manipulate information for them. That situated information in the environment is only harvested on a need-to-know basis. The last principle is that the environment is part of the cognitive system. Information is continuously relayed between the mind and world; thus studying the mind alone is not enough without the environment in which that mind lives.

The conceptualisation of abstract concepts, using "ideas and modes of reasoning grounded in the sensory-motor system", is called conceptual metaphor (Wilson, 2002). Research has shown that a mathematical activity which is embodied can result in a better understanding of abstract concepts (De Freitas & Sinclair, 2014). Studies have also shown that cognition can be strongly linked to and is actually "based in perception and action, and it is grounded in the physical environment" (Alibali & Nathan, 2012). This is the reason why this is a suitable theory for this study.

The last theoretical framework is RME. The development of RME was started by a Hans Freudenthal idea – that mathematics should be considered as a human activity (Freudenthal, 1991). RME is an instructional approach that aims at bridging the gap between abstract mathematical concepts and the real world. Mathematics is seen as a human activity and is therefore seen as connected to reality. Students are the actors of their learning and, guided by teachers and educators, enhance the development of the reinvention of the mathematical process (Gravemeijer, 1994).

In RME, contexts are important for developing mathematics understanding. These contexts provide a source for generating mathematical models, which in turn provide a deep understanding of where mathematics comes from (Van den Heuvel-Panhuizen, 2003). These contexts can be from the real world of the learner, from fiction or even from an area of mathematics that learners are already familiar with. When learners remain connected with the context, they are able to continue to learn conceptually and not only procedurally. They are able to learn without the need to simply memorise rules and procedures which are meaningless to them.

As noted by Van den Heuvel-Panhuizen and Drijvers (2014), the major characteristic of RME is that rich, 'realistic' situations from learners' contexts are given a prominent position in the learning process. They add that

> these situations serve as a source for initiating the development of mathematical concepts, tools, and procedures and as a context in which students can in a later stage apply their mathematical knowledge, which then gradually has become more formal and general and less context specific.
>
> (p. 521)

Approaching teaching and learning from the RME perspective helps learners to connect mathematical ideas and to rediscover concepts. This allows learners to formalise their informal understanding and intuition. Thus, in this chapter we regard providing learners with meaningful experiences through solving contextual issues they face every day, as the best way to teach mathematics for conceptual understanding.

Research Methodology

This chapter draws from work on outdoor mobile mathematics teaching and learning in South Africa (with a focus on pre-service teachers), Namibia (with a focus on learners) and Germany (with a focus on a modelling approach to learning). Data gathered from South Africa comprised work that was done with pre-service teachers at a university. Twenty-one pre-service teachers who were studying towards their Postgraduate Certificate in Education (PGCE) were purposively selected to take part in this study. They were introduced to the MCM app and to setting outdoor tasks. First, these students walked the trails their lecturer had created, then later they were asked to design their own tasks and to upload them onto the MCM app. They were interviewed after they had uploaded their tasks.

In Namibia, this study was conducted as a case study, in a school located in the Oshana region in the northern part of the country. A sample of 12 Grade 9 learners and their teacher were purposively selected on the basis that the Grade 9 teacher was willing to participate and had taken part in the MCM workshops. Only 12 learners were chosen because this was a pilot project which would later be expanded to the whole school and region. The teacher created outdoor mathematics tasks that were then uploaded on to the MCM app. The 12 learners were allocated to four groups of three. Only one group walked a trail at a time as this allowed the teacher to video record the proceedings fully while the learners worked on the tasks and interacted amongst themselves and with the MCM app on their smartphones. Interviews were conducted at the end of each trail. All the learners were interviewed by the teacher regarding their experiences during the trail.

For Germany, we draw from one study we have carried out during the last few years. In the study (Ludwig & Jablonski, 2019) the math trail was attempted by 153 ninth graders and 52 undergraduate university teacher students. We divided the students into groups of three, resulting in 51 groups of ninth graders and 17 groups of undergraduates. We set a mixed trail with different tasks about combinatorics, numbers, solid geometry and probability. The task we want to emphasise in this chapter was one that involved a flowerpot: *Calculate the volume of the flowerpot. Give the result in litres!* After the students completed the trail, we collected the smartphones and the students' notes. The shape of the flowerpot was a truncated cone. The formula for the volume of a truncated cone is not well known, so the participants had to come up with novel ways of calculating the volume. We categorised the students' solutions and found four categories of solutions for this task.

Findings

For the purposes of this chapter, we will only focus on a few of the findings from the vast amount of data that has been collected to date. The discussion in this section will be divided into three subsections dedicated to data from each of the countries we focused on in this chapter.

South Africa

The pre-service teachers who followed the trail created by their lecturer showed excitement and expressed amusement as they were following the trail. This was observed by the first author of this chapter as he walked the trails with them. Figure 9.4 shows some of the tasks that students had to solve outside the lecture room.

The tasks were from various domains of the South African school mathematics curriculum: functions/gradient (Figure 9.4a), measurement (Figure 9.4b) and number patterns (Figure 9.4c). One of the students, Pumza, mentioned that:

> … for the first time, I was able to make sense in real life of some of the mathematics that I have always learnt in the classroom. A good example for me

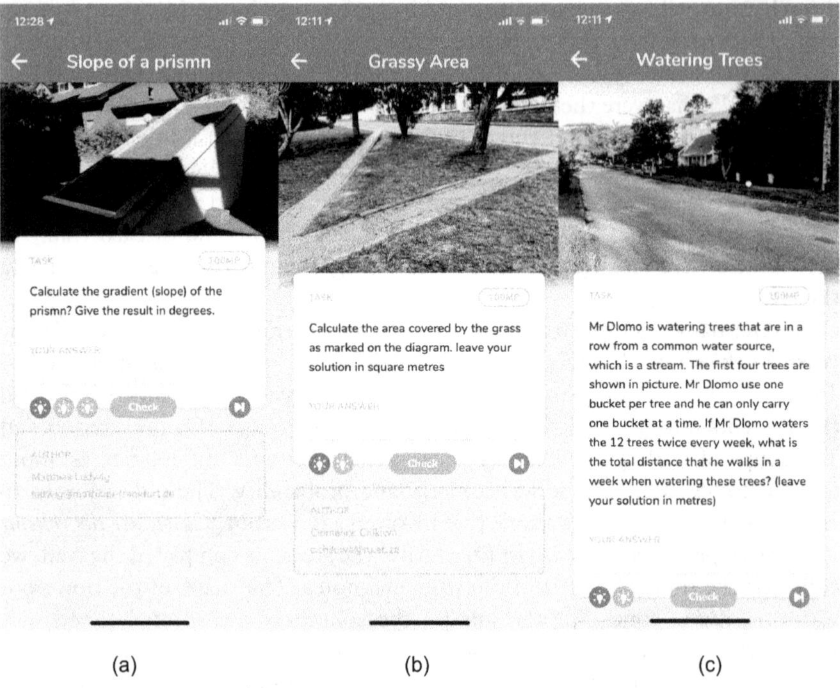

Figure 9.3 Three tasks from the South African trail.

> is that on Mr. Dlomo watering those trees. That is real and I never thought that involves patterns. This is an eye opener for me.

Buhle added:

> I have learnt the MCM makes mathematics real to a larger extent. Mathematics is made less abstract by dealing with actual objects that are familiar to you or even if they are not so familiar but because they are there physically.

The pre-service teachers reiterated the advantages of using outdoor real tasks. Molly commented on the availability of hints and solutions saying:

> I have noted that if I am to use this approach in my teaching in the future, it will save me from talking too much. Learners would just make use of hints if they are stuck and they can also check if their solutions are correct on the app. We experienced this as we were doing the tasks on this trail.

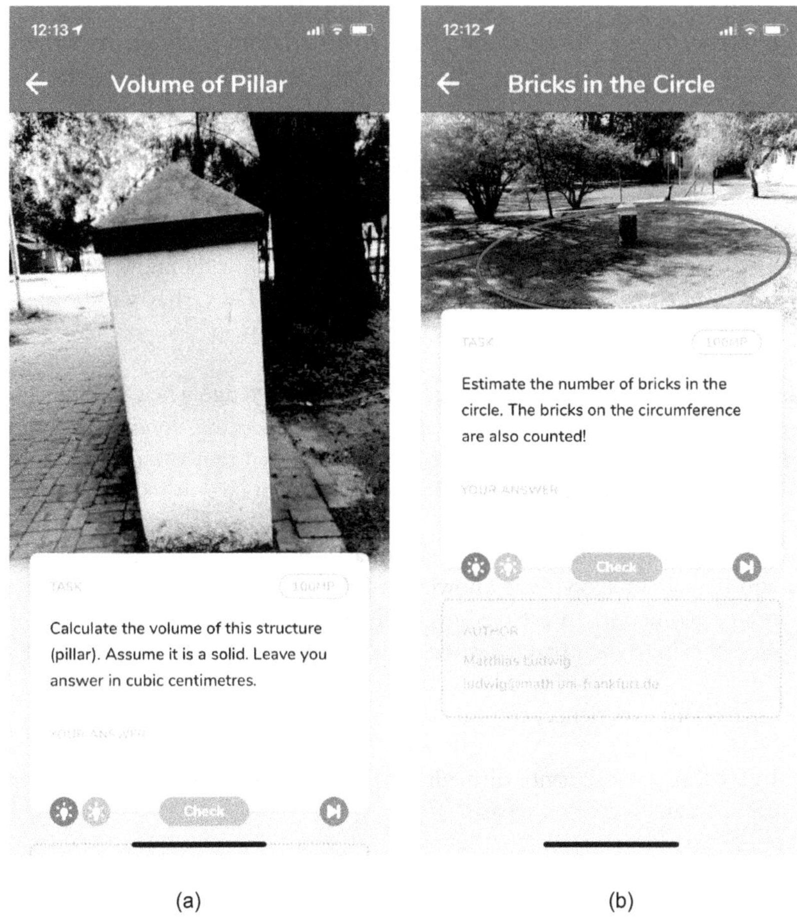

Figure 9.4 Examples of tasks created by the participating pre-service teachers.

In identifying the advantages of using math trails, the participating teachers confirmed the efficacy of these trails in their own contexts and environments.

After walking the trail that their lecturer had created, the pre-service teachers in this study were able to create a variety of tasks which they later uploaded on the MCM app. Some of them are shown in Figure 9.4.

The pre-service teachers' tasks were set at various cognitive levels, with most of them in the problem-solving and complex procedure levels (see Figure 9.4). Some of them, like Loyiso, acknowledged that it was not easy designing these tasks, saying,

> I learnt things that I had never thought of before. I have always been of the opinion that mathematics is best taught from the textbook and in the classroom but now I know I was wrong.

Thami commented and said,

> I have learnt how to use easily accessible materials to create maths activities. I now can rely on my learners' immediate environment outside the classroom to design tasks for them. Am sure they will be excited as I am.

Pumza gave one of the advantages of designing outdoor tasks and the MCM app:

> This helps learners to make sense of what they already know because the tasks I have created are around familiar objects. Thus, they will be able to take advantage of this familiarity during solving them.

This reiterates the importance of learners' prior knowledge when outdoor tasks from familiar environments are used, thereby promoting conceptual understanding of mathematical concepts. Molly commented that the use of outdoor tasks created within learners' environments has advantages. She said:

> They (learners) will know that mathematics is around them and hence they will be able to apply their knowledge from home. For example, if they are doing geometry and are dealing with tiles, they can also apply it at home and explain to their parents when they want to build something or put tiles.

Matt shared his experience, commenting:

> I have learnt that it is not difficult to apply maths in real life. It is even applicable on an everyday life basis.

Buhle added that in his recent experience after walking the trail and designing his own outdoor tasks,

> It actually appear[s] easier when mathematics is done outside than in the classroom because you physically see what you are dealing with and somehow it makes it easier to think.

The outdoor classroom provides opportunities for learners to interact with real and authentic objects, thereby making learning a living experience for them. The ability of learners to make judgements regarding the accuracy or inaccuracy of their solutions through comparing them with the actual object is important. Such a skill enables learners to gain two components of mathematical proficiency, referred to by Kilpatrick et al. (2001) as strategic competence and adaptive reasoning. Learners are thus empowered with the capacity for logical thought and reflection, and their procedures and solutions are affirmed.

While most pre-service teachers hailed the MCM app as a crucial tool to encourage learners to work outside, Loyiso gave one limitation he had experienced during the designing process. Loyiso argued that the MCM is suitable for some topics and less so with other topics:

> This MCM will only work best with some of the topics and not all maths topics in our curriculum. For example, it will not be easy to use MCM when teaching financial mathematics.

Namibia

The data from Namibia was on research with learners using the MCM app. It did not focus on the teaching. Data from observations revealed that there was much excitement as learners walked the trails. The learners shared roles and there was a great deal of interaction with the environment and amongst themselves.

The learners moved around a mopani tree in order to measure the circumference of the trunk. The size of the concrete block was smaller and so the learners did not need to physically move around it but used their fingers instead to discover the outlines as perimeters around the objects. In all instances, there was much physical activity during the problem-solving process. Sometimes they measured and remeasured the objects. Clearly, the learners were connected to the physical features of the objects. The use of actual objects in this case reiterates what Khan et al. (2015) articulated: "[A]a renewed study of visualisation in mathematics education must attend to the significant role of the body in space with other bodies" (p. 278).

The hints that were provided by the MCM application (see Figure 9.5) guided the learners on how to interact with the spatial relations of the objects they encountered on their trail.

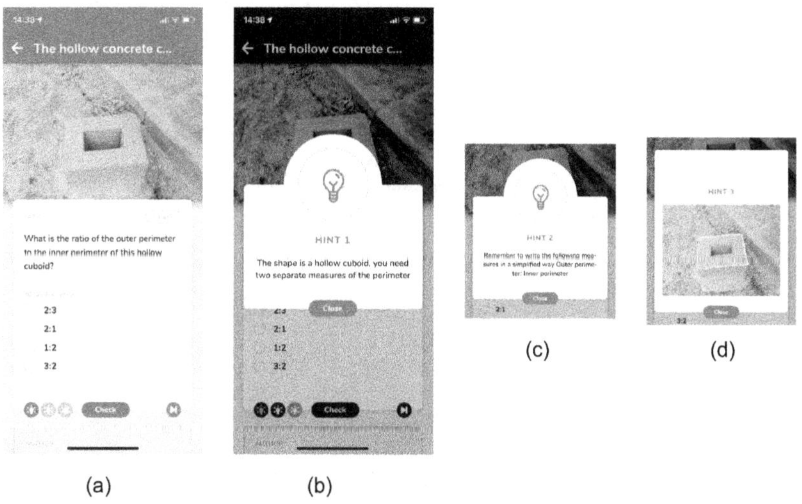

Figure 9.5 The hollow concrete block task and the hints provided by the app.

We noted that the hints played a number of roles: they prompted the learners to remember that the perimeter is a measure around a given object, they helped to motivate the learners when they could not find the solution to the problem and they helped the teacher talk far less during the learning process. The learners preferred to consult the hints, rather than ask the teacher. This fostered autonomy in the learners in that they had to make a decision to ask for a hint or not. Table 9.1 shows one of the conversations that transpired when the learners were working in their group using the hints that the MCM portal provided.

The stepped hints helped learners strengthen the way in which they processed their visualisations, as observed during their discussion in Table 9.1. With the school garden gate task (see Figure 9.6), the reading of the hint guided learners to the correct procedure and thus prevented the group from subtracting the length of the bars forming the rhombus from that of those forming the rectangle. Also, the content of the hints reminded learners of relevant prior knowledge – adding up all the lengths represented the perimeter for the bar. We noted that in many cases, learners were able to use their prior knowledge on the actual object they were working on. Figure 9.6 shows the school garden gate in addition to the sketch drawn by the learner to indicate that she already knew that a rhombus had four equal sides.

The learners' prior knowledge, together with the hint provided, was also used to deconstruct the complex shape of the gate into a simple square and a rhombus. This made it easy for learners to obtain separate measures when they subtracted the perimeter for the inner rhombus from the perimeter for the outer square. As shown in Figure 9.6, the scribe in the group made a sketch on the paper as they noted the rhombus on the gate. This resulted in this group being able to translate the actual concrete shape onto paper so that they could solve the problem. This ability to oscillate between the real object and its representation was also evident in many other problems that learners solved.

Table 9.1 Learners use stepped hints to improve their method

Learner	*What was said by learners*	*Researchers' observations*
L4O	Hint number one says: First, you need to calculate the total cost of the metal... now how can you use the metal length (to calculate the cost)?	L4O use the smartphone to read the first hint of the task
L5O	Ooo, so now you have to subtract ...6.12 minus 4.12?	The learner is inspired by the hint to make this statement
L6O	To subtract peni ano? Eenima atudhi addinga ... eyamukulo ndiyaka lyo 4.12 otatu li plus no 6.12 and then atukeli multiply nawa no 12.99 [What is there to subtract? We are adding the things (lengths) together ...on that length of 4.12 we add 6.12 then the answer we get, we multiply it with the 12.99 (which is the cost of the bar per metre)]	Out of excitement L5O quickly grabbed the scientific calculator from L6O in order to perform the alternative method for correcting the mistake that they made while making their first attempt on the solving of this task

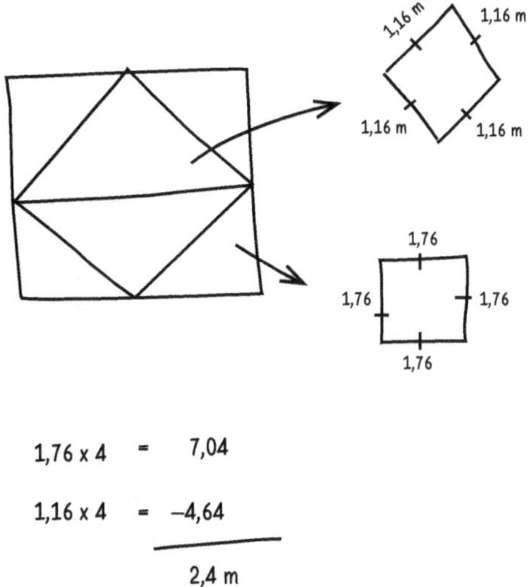

Figure 9.6 The school garden gate and its sketch.

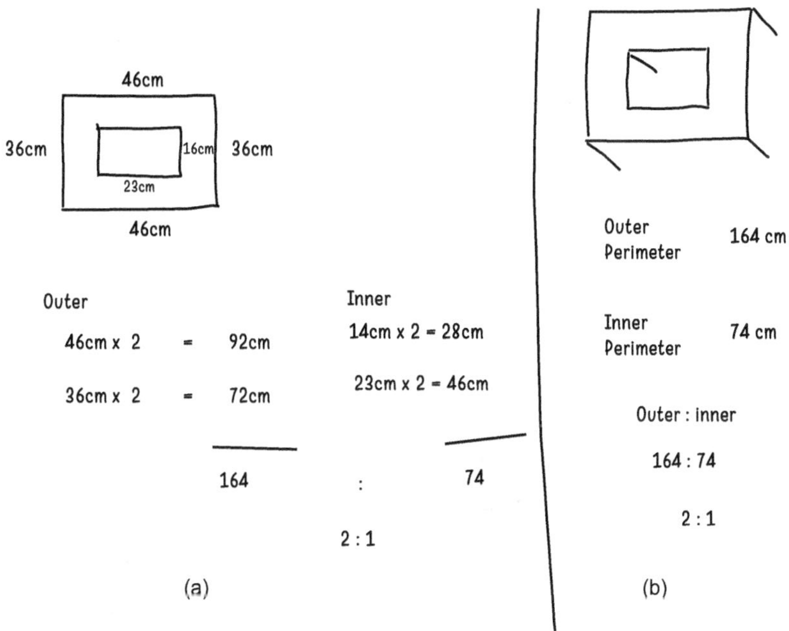

Figure 9.7a Group 2 calculations.
Figure 9.7b Group 3 calculations.

Figure 9.7 shows Group 2 and 3's solving of the problem. They, too, were also able to translate the actual object into a sketch. While each group approached the problem differently in terms of what they measured, they both managed to solve the task correctly.

Figure 9.7b shows how Group 3 used a method different from Group 2 when measuring length and width and applying the formula, during the solving of the hollow concrete cuboid task. The Group 3 learners determined the perimeter of the block by putting the measuring tape around the entire object, then they applied the same method to take a single measure for the inner perimeter of the hollow cuboid. Figure 9.7 illustrates how the two methods that were used varied and how both groups arrived at almost the same solutions.

We thus argue that working with authentic outdoor objects, where individuals have the freedom to use methods that work for them in their own situations, enhances learning.

We observed that the learners employed rich spatial perception when they came into physical contact with the objects involved in the tasks. This was noted in most of the tasks that learners were required to solve, including the pole task given as follows: *Calculate the volume of metal used to make the verandah pole leaving your answer in cm^3. Assume that the pole is solid.*

The sensory perception that learners employed as a result of the hint seemed to imprint certain images on L5O's mind – see Table 9.2. He folded his hand around the pole, which he perceived as a cylinder. Table 9.2 demonstrates this incident.

The data in Table 9.2 shows that reading aloud the first hint prompted L5O to intuitively use his sense of touch to perceive the pole as a physical artefact. This may be evidence that this learner's mental cognition was prompted into action because as soon as the task hint was read, learner L5O folded his fingers around the pole.

We noted that the holding of the pole helped the other learners to get an idea of what was required concerning the task at hand. For example, the way L5O closed his fingers around the pole and mentioned the word 'cylinder' could have suggested to other learners that the pole was indeed the shape of a cylinder and that they needed to identify the circumference as the first dimension that they needed to measure.

Table 9.2 An excerpt showing how learners perceive a pole as a cylinder

Learner	*What learner said*	*Researcher's observation*
L4O	The pole is a cylinder	L4O read the first hint of the task.
L5O	Pole…	Repeating part of the first hint for the task that was read by L4O tempted L5O to touch the pole by folding his hand around the pole as seen in Figure 9.8. The other group members focused on the pole which L5O was trying to acquaint himself with

His visual-spatial experience of the pole contributed to L5O's understanding of the pole as a cylinder. It became clear to Group 2 members that the circumference was needed in order to find the radius, which they were not able to measure directly. We argue that the act of touching the pole seemed to imprint the cylinder image in the learners' minds.

Germany

In Germany, outdoor mathematics is often associated with mathematical modelling. In this part of our chapter, we will try to show which partial competencies of modelling are enabled when working on MCM tasks. As already mentioned, most of the MCM tasks focus on simplifying and mathematising an authentic situation into an adequate mathematical model, which corresponds to steps 2 and 3 of Blum and Leiss (2007)'s seven-step modelling cycle (see Figure 9.2). In simplifying, important information is separated from unimportant information, which is then taken away from the authentic setting. Mathematisation involves the translation of the simplified setting into mathematical models (Greefrath et al., 2013). This is illustrated by the sample task in Figure 9.8.

In the flowerpot task shown in Figure 9.8, the object can be described fairly accurately as a truncated cone. Three different approaches were observed among the learners in a ninth grade class. The multitude of approaches can probably be attributed to the fact that the formula for the volume of a truncated cone was not familiar to every member of the group. The following modelling was performed (see Figure 9.9):

Here, each solution emphasises a different mathematical model as the observed learners solved the task mathematically in three different ways. The first solution variant in Figure 9.9a approximates the result via the mean value of the volume of a cylinder with the large radius (R) and a cylinder with the small radius (r); thus the learners used the formula:

$$V = \frac{R^2 + r^2}{2} \cdot h \cdot \pi$$

The second possible solution shown in Figure 9.9b approximates the result by a middle cylinder. For this, they took the mean value of the small (r) and large radius (R) as the radius, resulting in the following mathematical model:

$$V = \left(\frac{R + r}{2}\right)^2 \cdot h \cdot \pi$$

The third solution shown in Figure 9.9c is based on knowledge of the formula of the volume of a truncated cone:

$$V = \frac{R^2 + Rr + r^2}{3} \cdot h \cdot \pi$$

It is particularly interesting here that the end results of all three approaches hardly differ. On the one hand, this can be explained by the carefully collected

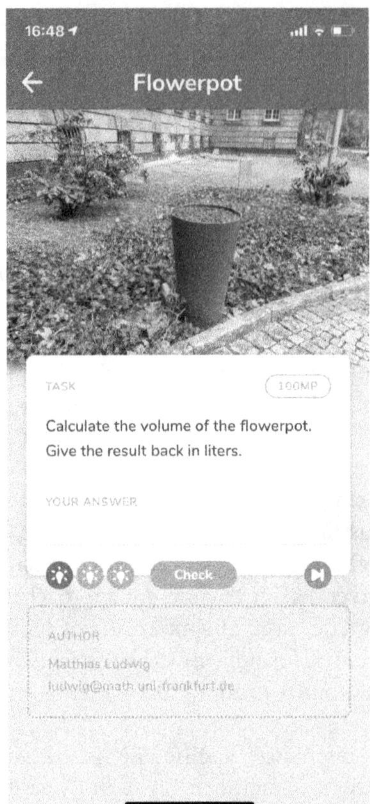

Figure 9.8 Flowerpot task.

measurements of all three groups of learners. On the other hand, the shape of the truncated cone shown here does not differ that much from an exact cylinder, so that the models used only led to minor deviations.

Each solution deals with different mathematical models. In each category, the learners had to create their own real model and then mathematise by adding variables they had to measure from the actual object. They had to think and decide on which data they needed to measure. This differed with the way that theoretical modelling tasks are usually presented to learners in the classroom. When we look at the numbers of learner groups we have worked with in Germany in the whole study, 20% of the ninth graders were not able to solve the task in an appropriate way. The first approach, that is, of using the mean of the radii (see Figure 9.9a), was often used (60%). Some learners used the first approach in another way: they measured the circumference of the truncated cone in the middle of its height and got the mean circumference at once. The second approach (calculating the mean of a big and small cylinder) was used by 15%

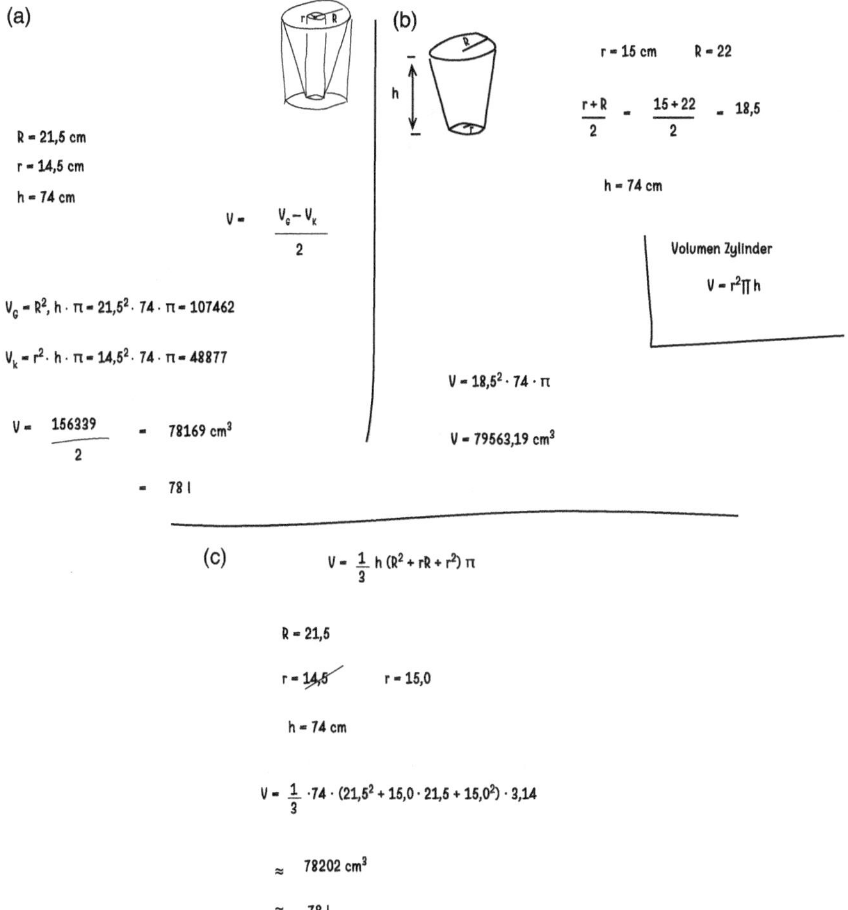

Figure 9.9 The three different learner solutions.

of the ninth graders. The third approach (the formula of the truncated cone) was seldom used. Overall, we observed that by collecting data from authentic settings, learners developed a strong connection to mathematical modelling (Blum & Leiss, 2007).

Conclusion

While our focus in this chapter was not to compare Grade 9 learners from these three countries, we noted that because of the differences in curricula being followed, in Germany, these types of problem solving are called modelling, which is not the case with Namibian and South African curricula. While modelling is not explicitly taught and emphasised in Namibia and South Africa at Grade 9 level, modelling ideas are incorporated. The use of outdoor tasks has the potential to encourage the understanding of elements of modelling in Grade 9 and also mathematics in general. The use of MCM tasks has a longer history in Germany than in Southern Africa; hence our focus was not on comparing these learners. In this chapter we thus argued that well-organised activities outside the classroom have the potential to contribute significantly to the quality and depth of children's mathematical learning. Outdoor math trails supply further evidence that such enhanced learning has the potential to further promote improved access to mathematical concepts. The outdoor tasks of the MCM have advantages of situatedness and hence they become meaningful, stimulating, challenging and exciting for learners. When these outdoor tasks in the form of mathematics trails are attempted on their own, or in conjunction with the use of smartphones, advantages associated with the use of visualisation processes in teaching and learning are tapped into. Most importantly, the mathematics trails provide an inclusive learning environment – thus providing great potential to enhance epistemological access for all learners. Outdoor tasks through mathematics trails invite all learners, irrespective of their classroom achievement level, to participate successfully in problem-solving activities and gain a sense of pride in the mathematics they create.

Note

1 According to data reported in https://www.geopoll.com/blog/mobile-penetration-south-africa/, in 2019 in South Africa "with a population of 58.56 million there were 101.9 million mobile subscribers across five main mobile network operators, resulting in the unique subscriber penetration rate of 170% (McCrocklin, 2021)."

References

Alibali, M. W., & Nathan, M. J. (2012). Embodiment in mathematics teaching and learning: Evidence from learners' and teachers' gestures. *Journal of the Learning Sciences, 21*(2), 247–286. https://doi.org/10.1080/10508406.2011.611446

Arcavi, A. (2003). The role of visual representations in the learning of mathematics. *Educational Studies in Mathematics, 52*(3), 215–241.

Barlovits, S., & Ludwig, M. (2021). MCM@HOME: A concept for teaching and learning mathematics during the corona crisis. *Proceedings of EDULEARN21 Conference 5th-6th July 2021*, 1588–1597.

Blane, D. C., & Clarke, D. (1984). *A Mathematics Trail Around the City of Melbourne*. Monash: Monash Mathematics Education Centre, Monash University.

Blum, W., & Leiss, D. (2007). How do Students and Teachers deal with Mathematical Modelling Problems? The example "Sugarloaf". In C. Haines, P. Galbraith, W. Blum, & S. Khan (Eds.), *Mathematical Modelling (ICTMA 12): Education, Engineering and Economics: Proceedings from the Twelfth International Conference on the Teaching of Mathematical Modelling and Applications* (pp. 222–231). Chister Horwood Publishing.

Blum, W., Galbraith, P.L., Henn, H.-W., & Niss, M. (2007). *Modelling and Applications in Mathematics Education*. New York: Springer.

Cahyono, A. N., Ludwig, M., & Marée, S. (2015). Designing mathematical outdoor tasks for the implementation of The MathCityMap-Project in Indonesia. In C. Vistro-Yu (Ed.), In pursuit of quality mathematics education for all. *Proceedings of the 7th ICMI-East Asia Regional Conference on Mathematics Education* (pp. 151–158). Quezon City: Philippine Council of Mathematics Teacher Educators (MATHTED), Inc.

Cahyono, A. N. (2018). *Learning Mathematics in a Mobile App-supported Math Trail Environment.* Heidelberg: Springer International Publishing.

Cohen, C. A., & Hegarty, M. (2007). Individual differences in use of external visualizations to perform an internal visualization task. *Applied Cognitive Psychology: The Official Journal of the Society for Applied Research in Memory and Cognition, 21*(6), 701–711.

Cohen, J. (2019). Math trails reveal the beauty of numbers and patterns in nature. https://greenschoolsnationalnetwork.org/math-trails-reveal-the-beauty-of-numbers-and-patterns-in-nature/

Daniel, S. J. (2020). Education and the COVID-19 pandemic. *Prospects*, 1–6. https://doi.org/10.1007/s11125-020-09464-3

De Freitas, E., & Sinclair, N. (2014). *Mathematics and the Body: Material Entanglements in the Classroom.* Cambridge: Cambridge University Press.

Duncan-Williams, K. (2020, April 19). *South Africa's Digital Divide Detrimental to the Youth.* Mail & Guardian. https://mg.co.za/article/2020-04-19-south-africas-digital-divide-detrimental-to-the-youth/

EduTech Tours. (2017). How South Africa leverages mobile learning to increase opportunities for all. https://medium.com/edtech-tours/how-south-africa-leverages-mobile-learning-to-increase-opportunities-for-all-74ec6b6d8f42

Fabian, K., Topping, K. J., & Barron, I. G. (2018). Using mobile technologies for mathematics: effects on student attitudes and achievement. *Educational Technology Research and Development, 66*(5), 1119–1139. https://psycnet.apa.org/doi/10.1007/s11423-018-9580-3

Freudenthal, H. (1991). *Revisiting Mathematics Education.* China lectures. Kluwer, Dordrecht.

Friedrich, H. F., & Mandl, H. (1992). Learning and thinking strategies – a problem breakdown. In F. Fischer, J. Bruhn, C. Gräsel, & H. Mandl (Eds.), *Fostering Collaborative Knowledge Construction with Visualization Tools* (pp. 213–232). Frankfurt: Learning and Instruction.

Gafoor, K. A., & Kurukkan, A. (2015). Why high school students feel mathematics difficult? An exploration of affective beliefs. *Paper presented in UGC Sponsored National Seminar on Pedagogy of Teacher Education – Trends and Challenges* at Farook Training College, Kozhikode, Kerala on 18th and 19th August 2015.

Gravemeijer, K. P. E. (1994). *Developing Realistic Mathematics Education.* Utrecht: CD-ß Press/Freudenthal Institute.

Greefrath, G., Kaiser, G., Blum, W., & Borromeo Ferri, R. (2013). Mathematisches Modellieren – eine Einführung in Theoretische und Didaktische Hintergründe. *Mathematisches Modellieren für Schule und Hochschule* (pp. 11–38). Wiesbaden: Springer Spektrum.

Gurjanow, I., Ludwig, M., & Zender, J. (2017). What influences in-service and student teachers use of MathCityMap? *CERME 10*, Dublin, Ireland.

Gutiérrez, A. (1996). Visualization in 3-dimensional geometry: In search of a framework. In L. Puig & A. Guttierez (Eds.), *Proceedings of the 20th Conference of the International Group for the Psychology of Mathematics Education, University of Valencia, Valencia, 1* (pp. 3–19).

Hayes, B., & Scassellati, B. (2013). Challenges in shared environment human-robot collaboration. In *Proceedings of the 8th ACM/IEEE International Conference on Human Robot Interaction (HRI 2013)*. Workshop on Collaborative Manipulation.

Isenberg, P., Elmqvist, N., Scholtz, J., Cernea, D., Ma, K-L., & Hagen, H. (2011). Collaborative visualization: definition, challenges, and research agenda. *Information Visualization, 10*(4), 310–326. https://doi.org/0.1177/1473871611412817

Kemmerer, D., Miller, L., MacPherson, M. K., Huber, J., & Tranel, D. (2013). An investigation of semantic similarity judgments about action and non-action verbs in Parkinson's disease: Implications for the embodied cognition framework. *Frontiers in Human Neuroscience, 7*(146). https://doi.org/10.3389/fnhum.2013.00146

Khan, S., Francis, K., & Davis, B. (2015). Accumulation of experience in a vast number of cases: Enactivism as a fit framework for the study of spatial reasoning in mathematics education. *ZDM, 47*, 269–279.

Kilpatrick, J., Swafford, J., & Findell, B. (Eds.). (2001). *Adding it Up: Helping Children Learn Mathematics*. Mathematics Learning Study Committee. Washington, DC: National Academy Press.

Kukulska-Hulme, A., Lee, H., & Norris, L. (2017). Mobile learning revolution: implications for language pedagogy. In C. A. Chapelle, S., & Sauro (Eds.), *The Handbook of Technology and Second Language Teaching and Learning* (pp. 217–233). Oxford: Wiley & Sons.

Lakoff, G., & Núñez, R. E. (2000). *Where Mathematics comes from: How the Embodied Mind Brings Mathematics Into Being*. New York: Basic Books.

Langoban, M. A. (2020). What makes mathematics difficult as a subject for most students in higher education? *International Journal of English and Education, 9*(3), 214–220.

Ludwig, M., & Jablonski, S. (2019). *Doing Math Modelling Outdoors. A Special Math Class Activity Designed with MathCityMap. 5th International Conference on Higher Education Advances* (pp. 901–909). Valencia, Spain: HEAd'19. Universitat Politecnica de Valencia, Valencia.

Ludwig, M., & Jesberg, J. (2015). Using mobile technology to provide outdoor modelling tasks – The MathCityMap-Project. *Procedia – Social and Behavioral Sciences, 191*(1), 2776–2781.

Makina, A. (2010). The role of visualization in developing critical thinking in mathematics. *Perspectives in Education, 28*, 24–33.

Mariotti, M. A. (2009). Artifacts and signs after a vygotskian perspective: The role of the teacher. *ZDM Mathematics Education, 41*, 427–440. https://doi.org/10.1007/s11858-009-0199-z

McCrocklin, S. (2021). Mobile Penetration in South Africa. https://www.geopoll.com/blog/mobile-penetration-south-africa/

McDonald, S., & Watson, A. (2010). *What's in a Task: Generating Rich Mathematical Activity*. Oxford: QCDA.

Medienpädagogischer Forschungsverbund Südwest (MPFS). (2020). *JIM-Studie, Jugend, Information, Medien*. Die Medienanstalten für Baden- Württemberg und Rheoinland-Pfalz. https://www.mpfs.de/fileadmin/files/Studien/JIM/2020/JIM-Studie-2020_Web_final.pdf

Namibia Press Agency. (2007, January 18). No Cell Phones in Class. The Namibian. https://www.namibian.com.na/index.php?id=32776&page=archive-read

Nardi, E. (2014). Reflections on visualization in mathematics and in mathematics education. In M. Fried (Ed.), *Mathematics & Mathematics Education: Searching for Common Ground* (pp. 193–220). Springer: Dordrecht.

Ngesi, N., Landa, N., Madikiza, N., Cekiso, M. P., Tshotsho, B. & Walters, L. M. (2018) Use of mobile phones as supplementary teaching and learning tools to learners in South Africa, *Reading & Writing*, *9*(1), a190. https://doi.org/10.4102/rw.v9i1.190

Núñez, R. E., Edwards, L. D., & Filipe Matos, J. (1999). Embodied cognition as grounding for situatedness and context in mathematics education. *Educational Studies in Mathematics*, *39*, 45–65. https://doi.org/10.1023/A:1003759711966

Piaget, J. (1952). *The Origins of Intelligence in Children*. New York: W.W. Norton & Co.

Porter, G., Hampshire, K., Abane, A., Munthali, A., Robson, E., Mashiri, M., & Tanle, A. (2012). Youth, mobility and mobile phones in Africa: Findings from a three-country study. *Journal of Information Technology for Development*, *18*(2), 145–162. https://doi.org/10.1080/02681102.2011.643210

Shapiro, L. (Ed.). (2014). *The Routledge Handbook of Embodied Cognition*. London and New York: Routledge.

Sharples, M., Taylor, J., & Vavoula, G. (2010). A theory of learning for the mobile age. In *Medienbildung in Neuen Kulturräumen* (pp. 87–99). VS Verlag für Sozialwissenschaften.

Statista. (2021). Number of Mobile Cellular Subscriptions per 100 Inhabitants in Namibia from 2000 to 2019. https://www.statista.com/statistics/509593/mobile-cellular-subscriptions-per-100-inhabitants-in-namibia/

Trafton, P. R., Reys, B. J., & Wasman, D. G. (2001). Standards-based mathematics curriculum materials: A phrase in search of a definition. *Phi Delta Kappan*, *83*(3), 259–264.

UNICEF. (2016). *Exploratory Research Study on Knowledge, Attitudes and Practices of ICT Use and Online Safety Risks by Children in Namibia*. UNICEF Namibia. https://www.unicef.org/namibia/na.COP_Research_Report_2016.compressed.pdf

Van den Heuvel-Panhuizen, M. (2003). The didactical use of models in realistic mathematics education: An example from a longitudinal trajectory on percentage. *Educational Studies in Mathematics*, *54*(1), 9–35. http://dx.doi.org/10.1023/B:EDUC.0000005212.03219.dc

Van den Heuvel-Panhuizen, M., & Drijvers, P. (2014). Realistic mathematics education. In S. Lerman (Ed.), *Encyclopedia of Mathematics Education* (pp. 521–525). Dordrecht: Springer.

Vygotsky, L. (1978). *Mind in Society*. Cambridge, MA: Harvard University Press.

Wilson, M. (2002). Six views of embodied cognition. *Psychonomic Bulletin & Review*, *9*(4), 625–36. https://doi.org/10.3758/bf03196322. P

Zender, J., & Ludwig, M. (2019). The long-term effects of Mathcitymap on the performance of German 15-year-old students concerning cylindric tasks. In U. T. Jankvist, M. van den Heuvel-Panhuizen, & M. Veldhuis (Eds.), *Proceedings of the Eleventh Congress of the European Society for Research in Mathematics Education*. Utrecht, The Netherlands: Freudenthal Group & Freudenthal Institute, Utrecht University and ERME.

Part 4

Visualisation, Semiotics and Language

10 Visualisation and Semiotics

A Case for Gestures

Cristina Sabena and Marc Schäfer

> Gestures make my class interesting ... help learners understand concepts more easily ... help to illustrate and visualise mathematical ideas... help in getting learners to participate in class. They assist in emphasising and clarifying what I teach. Gesturing particularly helps learners with learning difficulties. I believe it improves performance. Some learners react better to gestures than others. Some learners prefer when I only talk. I believe that gesturing is an inborn thing. Some teachers gesture a lot – some don't. Gesturing improves the quality of the lesson. Using gestures is a form of communication.
>
> (Responses of research participants in Namibia)

Introduction

As long ago as the 1990s Hans Freudenthal drew attention to *processes* as a crucial issue in mathematics education research: "... the use of and the emphasis on *processes* is a *didactical principle*" (Freudenthal, 1991, p. 87, emphasis as in the original). Considering in particular how teaching-learning processes develop in the classroom, every observer (and teachers as privileged observers in the classroom) may have noted that students and teachers activate a variety of resources to carry out their actions and to communicate with each other: words (spoken or written), written representations, extra-linguistic ways of expression (gestures, other bodily actions and gazes), tools and so on. In Vygotskian-oriented approaches, such resources are considered to play an important role not only in communication, but also in the cognitive processes underpinning the production of mathematical meaning (Arzarello et al., 2009; Radford, 2009).

An important push to reconsider the role of the body in mathematical thinking has been provided by numerous embodied cognition studies. At the turn of the millennium, the provocative essay *Where Mathematics Comes From* by George Lakoff and Rafael Núñez (Lakoff & Núñez, 2000), for example, pointed out the crucial role of perceptual and bodily aspects on the formation of abstract concepts, including mathematical ones. The publication of this volume opened up a rich debate in mathematics education at the time (see e.g. Research Forums at PME conferences) and prompted a reflection about the cognitive roots of mathematical ideas (e.g. see Tall, 2000).

DOI: 10.4324/9781003172420-14

Criticising platonic idealism and the Cartesian mind-body dualism, Lakoff and Núñez (2000) advocated that all kinds of ideas – including the most sophisticated mathematical ones – —are founded on our bodily experiences and develop through metaphorical mechanisms. Their account brought about two claims: first, our understanding of concepts is structured by our bodies and our everyday functioning in the world; second, human beings come to understand and know abstract concepts (including mathematics) through their sensory-motor experiences, which are largely metaphorical in nature. This is not a completely new idea, if we think of the psychological studies by Piaget, Vygotsky, and Montessori, and the philosophical ones by Husserl, Merleau-Ponty, within the phenomenological discourse. What was new in the notion of embodiment, and constituted a constructive innovation beyond the radical criticism that characterised publications at the time, was the theorisation about the cognitive mechanism through which mathematical ideas were linked to everyday concepts, that is, the *cognitive metaphor* (e.g. time conceptualised as space, and sets conceptualised as containers).

This idea appeared to receive confirmation by later neuroscientific results related to the discovery of the so-called 'mirror neurons', which fire when a subject does a certain action (e.g. grasping an object), as well as when he/she sees somebody else doing the same action (Gallese & Goldman, 1998; Rizzolatti & Gallese, 1997). Other neurons are found to be 'multimodal', firing when a subject does an action, or sees it, or hears its noise.

The term 'multimodality' thus emerged in the neuroscientific field to indicate a feature of human cognition, opposed to 'modularity', stressing a deep integration between aspects that were traditionally considered as neuronally separated, such as action and perception. Furthermore, such integration is crucial not only for motor control, but also for planning actions, an imperative part of *thinking*.

Finally, in the communication field, the term 'multimodal' is used with reference to multiple modalities that we can exploit in order to communicate and express meanings to those we engage with: words, sounds, images and so on (Kress, 2004). Currently, these dimensions are acquiring increasing attention due to the diffusion of new technological opportunities.

Drawing on the results arising from these different research fields, the term 'multimodality' has emerged in mathematics education to refer to the relevance and mutual co-existence of a variety of cognitive, material and perceptive modalities or resources in mathematics teaching-learning processes, and more generally in mathematical thinking:

> These resources or modalities include both oral and written symbolic communication as well as drawing, gesture, the manipulation of physical and electronic artifacts, and various kinds of bodily motion.
>
> (Radford et al., 2009, pp. 91, 92)

An emerging component of multimodality in the classroom is the role of gestures in synergy with spoken language and written inscriptions. In this chapter we draw on a multimodal perspective and present three studies from Italian and Namibian contexts, to show how gestures may have relevant roles in the mathematics teaching-learning process in the classroom. Before presenting our empirical-analytical section, we briefly summarise key results of gesture studies and introduce a semiotic perspective that frames them in a mathematical activity context. Finally, we discuss the connection between visualising and visualisation.

Gestures in Communication and Thinking Processes

Gesturing is a widespread phenomenon found in all cultures. It occurs across a wide range of tasks and ages, even when the listener is not physically present – when using the telephone (McNeill, 1992) and in congenitally blind people's conversations (Iverson & Goldin-Meadow, 2001). Gestures are part of what is called 'nonverbal communication', which includes a wide-ranging array of behaviours such as the distance between people in conversation, eye contact, voice, body posture and so on. In the traditional view of communication, all these acts (although important in framing a conversation) often have little to do with the conversation itself, which is implied in the uttered speech (the 'verbal communication'). Gestures are thus often interpreted as nothing more than handwaving, embellishments, a means outpouring an excess of energy, bids for the listener's attention or regulators in the communicative exchange.

After pioneering publications on gestures appeared in the 1980s, such as those by Kendon (1980), there was a proliferation of studies on gestures from the 1990s onwards, bringing to the fore interesting and relevant issues regarding the role of gestures in communicating, thinking, reasoning and learning.

Spontaneous gestures, particularly in conversational settings, received special attention in psychology, anthropology and cognitive linguistics, showing that gestures have a fundamental importance not only in communication, but also in *cognition*. In this context, the studies led by David McNeill have become classical within the field and also a reference point for those interested in gestures within mathematics education. McNeill and collaborators, focusing in particular on conversational settings, considered gestures as "the movements of the hands and arms that we see when people talk" (McNeill, 1992, p. 1), and, indeed, following McNeill's work on gestures, arms and hands, became the main focus of attention.

McNeill investigated how gestures and speech are intricately integrated with each other in conveying meanings in a particular discourse (McNeill, 1992, 2005). In order to describe the diversity of gestures, he introduced a classification consisting of four categories: iconic, metaphoric, deictic and beat (McNeill, 1992).

- *Iconic*: gestures that depict images of concrete objects or actions; they have a relationship of similarity with the semantic content of the speech they accompany.
- *Metaphoric*: gestures that depict images of abstract ideas, through a metaphoric use of form or space.

- *Deictics* (also called *pointings*): usually made with the index finger, they indicate objects or positions in space.
- *Beats*: by beating time, they provide a temporal highlight to speech.

McNeill's classification is not necessarily based on the physical characteristics of the gesture, but rather on the relationship with its meaning. Meaning, therefore, is not a feature inherent to the gesture itself, but becomes significant through interpretation against the background of the discourse (i.e. the co-occurring words), its context and the cultural backdrop of the gesturer and the discourse. Moreover, McNeill later pointed out that the same gesture could belong to more than one of the four categories above. He stressed that these categories are to be conceived as *dimensions* that characterise a gesture, rather than placings in a discrete categorical sense (McNeill, 2005).

Iconic, metaphoric and deictic gestures are also called 'representational' (Kendon, 1988; Kita, 2000), as they represent some aspects of the objects, either concrete or abstract, to which they refer. Metaphoric gestures give an image to something invisible, an abstract concept, as in the example of the 'conduit gesture' in English-speaking culture, in which the idea of a topic (which is being discussed in a conversation) is presented as a hand-held container. In other cultures, knowledge may be represented differently in gestures, for instance, by something the hand releases into the air (McNeill, 2005). This shows that metaphoric gestures incorporate cultural beliefs (McNeill, 2005).

Deictic gestures or pointings are movements of indication, usually (but not exclusively) made with the index finger, with the purpose of fixing the attention to a common reference, depending on the context of the verbal interaction. Although this gesture appears simple and self-explanatory, pointing can be a complex act (Kita, 2000). Pointing can be concrete or abstract. It is concrete if directed at an object which is actually present in space, and abstract if it is directed towards an empty space. Interestingly, concrete pointing is one of the first gestures to be developed in children, at about ten months of age (McNeill, 1992; Volterra et al., 2005). In the early stages of language development, concrete pointing is used in combination with language to name objects. Vygotsky's discussion of the origin of a child's pointing gesture as a sign of communication is famous. It illustrates the process of internalisation, starting from the meaning attributed by the mother to the actual movement of her child:

> The child attempts to grasp an object placed beyond his reach; his hands, stretched toward that object, remain poised in the air. His fingers make grasping movements. At this initial state pointing is represented by the child's movement, which seems to be pointing to an object – that and nothing more. When the mother comes to the child's aid and realises his movement indicates something, the situation changes fundamentally. Pointing becomes a gesture for others. The child's unsuccessful attempt engenders a reaction not from the object he seeks but from another person.

> Consequently, the primary meaning of that unsuccessful grasping movement is established by others.
>
> (Vygotsky, 1931/1978, p. 56)

Abstract pointing develops much later, usually from the age of 12. In abstract pointing, the gesture does not indicate any object or place, but generally points to an empty space. McNeill (1992) recognises a certain conceptual meaning to such pointed-at empty spaces, through a metaphorical use of space, in which meanings are given a spatial form or location.

In McNeill's earlier works (McNeill, 1992), the category of *cohesive gestures* was included, to indicate cases in which gestures tie together thematically connected but temporally distant parts of speech. More recently, the cohesive function of gestures has been further investigated by McNeill and theorised with regard to the notion of *catchment* (McNeill et al., 2001; McNeill, 2005). A catchment is recognisable when some characteristic of a gesture is repeated in at least two gestures, not necessarily consecutively: for example, the form of the gesture could be repeated, as well as the position in space of a deictic gesture. According to McNeill, a catchment indicates discursive cohesion and is due to the recurrence of visuo-spatial imagery in the speaker.

Looking at the semantic relationship between words and gestures, Goldin-Meadow and colleagues distinguished those cases in which gestures and words convey the same meaning (*'gesture-speech matches'*, e.g. when saying "so high" one makes a vertical gesture) from those cases in which gestures and words convey different (not necessarily contradictive but also complementary) meanings (*'gesture-speech mismatches'*, e.g. making a horizontal gesture when speaking of height) (Alibali et al., 2000; Goldin-Meadow, 2000, 2003).

Cases of special interest are the mismatching ones. These are gestures which reveal thoughts that are distinct from those revealed in the accompanied speech. In this respect, Goldin-Meadow's hypothesis is the following:

> A speaker who has produced a gesture-speech mismatch knows (at some level) the information conveyed in both modalities. However, the speaker has not yet developed a framework – either a knowledge framework over developmental time or a discourse framework over conversational time – within which those pieces can be fitted together.
>
> (Goldin-Meadow, 2003, p. 29)

In particular, mismatches have been found when, during problem-solving contexts, strategy decisions have to be made and different alternatives considered in formulating a plan of action. Garber and Goldin-Meadow (2002) experienced this when working with selected children and adults who engaged in the Tower of Hanoi game. In the classical conservation tasks of Piagetian tradition, mismatches are interpreted to indicate cognitive variability: children who produce a larger number of mismatches would be more ready to learn than those showing few mismatches on the same task (Goldin-Meadow, 2000). In other words, frequent mismatches between gesture and speech in children may indicate a heightened receptivity to instruction. This thesis has been confirmed by numerous studies carried out by means of interviews between an instructor and an interviewed child (Goldin-Meadow et al., 1999; Goldin-Meadow, 2003, 2006; Özçaliskan & Goldin-Meadow, 2009). Goldin-Meadow (2000) claims that teachers can therefore capitalise from a student's mismatching gestures, since they can gain information about their Zone of Proximal Development (Vygotsky, 1978):

> When a student's gestures convey information that is different from the information found in speech, those gestures can inform the teacher of thoughts that the student has but cannot (or at least does not) express in speech. Gesture may be one of the best ways that teachers have of discovering thoughts that are on the edge of a student's competence – what L.S. Vygotsky called the child's "zone of proximal development" (the set of skills a child is actively engaged in developing).
>
> (p. 104)

More generally, from a purely cognitive point of view, the benefit gestures bring is often interpreted in terms of *lightening the cognitive burden* in an analogue way that writing a problem down can reduce the effort needed to solve the problem. Gestures can therefore constitute an embodied form for off-loading cognitive work into the environment (Wilson, 2002).

Semiotic Approaches to Multimodality

Attention to multimodal aspects of mathematics learning has been given from different perspectives (Edwards, 2009; Nemirovksy, 2003; Radford, 2009; Roth, 2009). Some studies have adopted a semiotic lens and have considered gestures and embodied resources as part of the range of semiotic resources that students and teachers can exploit in their mathematical activities. They mostly refer to Radford's Objectification Theory (2009), which strongly influenced Arzarello's *Semiotic Bundle* lens (Arzarello, 2006), as a system made of the different *signs* and of their *mutual relationships*. These are typically produced by students and possibly their teacher, in classroom activities. For the purpose of this chapter, we will focus only on Arzarello's Semiotic Bundle. More precisely, a Semiotic Bundle is defined as:

a A set of signs which may possibly be produced with different actions that have an intentional character, such as uttering, speaking, writing, drawing, gesticulating or handling an artefact.
b A set of modes for producing signs and possibly transforming them; such modes can possibly be rules or algorithms but can also be more flexible action or production modes used by the subject.
c A set of relationships among these signs and their meanings embodied in an underlying meaning structure (Arzarello, 2006, p. 8).

In this definition, 'sign' is intended inclusively, after Vygotsky's and Peirce's theories. As a matter of fact, for both Vygotsky and Peirce there is no limitation upon what can be conceived as sign. Peirce, in his representational account of sign, defined it essentially as a relationship between a representamen (something that represents something else), an object (which is represented) and an interpretant (another representation referring to the same object, which starts the semiotic process): "A sign, or representamen, is something which stands to somebody for something in some respect or capacity" (Peirce, 1958).

This definition does not prescribe what can be considered as a sign: anything that enters into a process of semiosis can be considered a sign. On the contrary, other semiotic approaches considered in mathematics education place strong limitations on what can be considered as part of the semiotic processes: see, for instance, Duval's notion of 'semiotic register of representation', which has precise and codifiable rules of production and transformation (Duval, 2006). Also, Peirce's specification of signs in the three main categories of icon, index and symbol seems to point to the complexity of semiotic processes that have been analysed deeply with respect to mathematics epistemology (Otte, 2006).

Also, in the Vygotskian's functional perspective on signs, there are no *a priori* constraints on what can be considered as a sign. Signs are tools of reflection that allow individuals to plan action and a supplementary means to organise their behaviour:

> the invention and use of signs as auxiliary means of solving a given psychological problem (to remember, compare something, report, choose and so on) is analogous to the invention of tools in one psychological respect. The signs act as instrument of psychological activity in a manner analogous to the role of a tool in labor.
>
> (Vygotsky, 1931/78, p. 52)

In the Semiotic Bundle, the notion of a *semiotic set* is defined by broadening Ernest's definition of semiotic system (Ernest, 2006), with the purpose of including embodied features (and their idiosyncratic modes of production), as well as the classical semiotic registers (e.g. registers *à la* Duval). The word 'set' is used in the mathematical sense of 'collection of tokens'. In Semiotic Bundle studies, the expression 'semiotic resources' is often used as synonymous with the word 'sign', to underline something that is at the students' and teacher's disposal in the mathematics classroom. An example would be the students' words, gestures and drawn figures while solving a geometrical problem: all are different signs that constitute a semiotic set; they are not only present in the same problem-solving activity, but there may be some relationships between them. For example, a gesture may give the origin to a written diagram, showing a genetic relationship within the Semiotic Bundle (Sabena et al., 2012).

From a structural point of view, the Semiotic Bundle is characterised by two key features:

- its systemic character, referring to the relationships between the various kinds of signs (which are also called 'semiotic sets') at a certain moment (like a sort of 'semiotic picture') – *synchronic analysis*; and
- its dynamic nature, focusing on the evolution of signs and of their transformations through time (a sort of 'semiotic movie') – *diachronic analysis.*

Also, in gesture studies we find elements referring to a *synchronic analysis*, such as the speech-gesture relationship at a certain moment, and to a *diachronic analysis*, such as the repetition of a certain gesture over a long time, called 'catchment' (McNeill, 2005). And, like studies in the gesture field, it is essential to take the context into account in order to understand the contribution of a single semiotic resource within the Semiotic Bundle.

Looking at the mathematics teaching-learning processes in the classroom within a multimodal perspective has led to the identification of a new didactic phenomenon called the '*teacher's semiotic game*' (Arzarello et al., 2009). A semiotic game may occur when the teacher interacts with the students, for example, in classroom discussions or during group work. In a semiotic game, the teacher tunes in to the students' semiotic resources (e.g. words and gestures), and uses them to make the mathematical knowledge evolve towards scientifically shared meanings. For instance, the teacher may repeat a gesture that one student has used and accompany it with an appropriate verbal explanation. Semiotic games may contribute to the students' process of appropriating the culturally shared meaning of signs. The teacher should be aware of the importance of considering the multimodal character of the classroom activity:

> [semiotic games] allow the teacher to become suitably in tune with students' languages and, conversely, they allow students to achieve resonance with the teacher's languages and, through them, with the institutional knowledge. In order that such opportunities can be concretely realised, the teacher must be aware of the role that multimodality and semiotic games can play in teaching. Awareness is necessary for designing the conditions that foster positive learning experiences and for adapting her/his intervention techniques to the specific didactic activity.
>
> (pp. 107, 108)

Visualisation, Gestures and Embodiment

We argue in this chapter that, as much as gesturing is integral to communicating and thinking, it is also an integral aspect of visualising. Alibali (2005, p. 2) observes that "gestures are particularly good at expressing spatial and motor information". As gestures are spatial in nature, they are observable and, by implication, they are visual – they can be seen by an observer and can be used as a visual mediator in the various roles referred to above. Gestures are a powerful means to represent and communicate what Goldin-Meadow (2015) identifies as a world that is not present. They can thus be used to mediate the visualisation of abstract mathematical ideas or concepts. Castellon and Enyedy (2006) argue that when gestures are used in combination with speech, they can form important visual tools. This was the central focus of Chikiwa and Schäfer's (2020) research when they observed how gestures formed an integral and important component of selected mathematics teachers' communication store when teaching in a multilingual context. Visualisation in conjunction with gesturing are integral elements of the Semiotic Bundle and its multimodal resourcefulness. Visualisation in this context is seen as a complex interaction between product and a process – what Nemirovsky and Nobel (1997) describe as back-and-forth travel between external representations and mental representations. Arcavi (2003) recognises this intricate relationship when he defines visualisation as:

> the ability, the process and the product of creation, interpretation, use of and reflection upon pictures, images, diagrams, in our minds, on paper or with technological tools, with the purpose of depicting and communicating information, thinking about and developing previously unknown ideas and advancing understandings.
>
> (p. 217)

We, however, argue further that bodily actions such as gesturing are also integral components of Arcavi's process and product definition of visualisation and are an important aspect of cognition. According to Di Paolo et al. (2007), cognition is a an embodied action grounded in bodily actions. An embodied view of cognition, according to Antle et al. (2008), grants the body situatedness in the environment, which in turn shapes the mind. The intricate interlocking of the body, mind and environment is a central tenet of enactivism which asserts that the mind, body and environment are indivisible. Varela et al. (1991) argue that the individual and the environment must be considered together, that one cannot separate knowledge from doing and from the body and that knowing is doing – which in the end is inseparable from self-identity or being. Gestures (bodily actions) thus cannot be separated and isolated from the identity of the gesturing individual.

Some Empirical Evidence from Namibia and Italy

The following presents evidence from three research studies: two from Namibia and one from Italy. The former are descriptive accounts of gestures used by selected teachers, whereas the latter is an analytic account of how a selected teacher and her learners used gestures in a problem-solving context.

Examples from Namibia: Gestures as Visualisation Tools

In two recent small-scale, mixed method case studies in Namibia (Namakalu, 2018; Haipinge, 2021) two researchers spent time with selected teachers observing the types of gestures that they would typically use over a period of time when teaching mathematics. Both studies were underpinned by an enactivist and embodied cognition theoretical framework and the participant teachers were purposefully selected. The criteria for selection included a willingness to be observed and video recorded, to reflect on their teaching and share their views and perceptions on their practice. Namakalu's (2018) study involved three purposefully selected junior primary mathematics teachers at a school in northern Namibia. The study sought to describe and understand how gestures were used to accompany mathematical instruction, with a specific focus on their role as visualisation tools (Schäfer & Namakalu, 2019) with particular reference to McNeill's (1992) taxonomy of gestures. The selected teachers were each observed ten times over a period of one term. They were then each interviewed in a stimulus-recall interview where the researcher was interested in understanding the teachers' own views on the visualisation roles of their gestures in their teaching. As Figure 10.1 shows, the most frequently used gesture in the 2018 study was the pointing gesture, followed by the beat gesture.

In the interviews all three teachers related how they used gestures and commented on the significant role that they played in their teaching – see Table 10.1.

In this study, symbolic gestures embody a precise, culturally shaped meaning, conveyed without the use of words. These kinds of gestures, which Kendon called emblems, are not included in McNeill's framework, which is restricted to spontaneous gestures accompanying speech (for a deeper discussion, see McNeill, 1992).

As Schäfer and Namakalu (2019) noted in this 2018 study, there was evidence that the use of gestures provided for a visually rich learning environment. The participating teachers agreed that gestures facilitated and strengthened both instructional and conceptual communication. Schäfer and Namakalu (2019) concluded that it is thus important that the appropriate use of gestures is recognised as a legitimate teaching strategy that supports good teaching and they argue that, although many gestures are uttered spontaneously, for them to be meaningful, they should be used strategically.

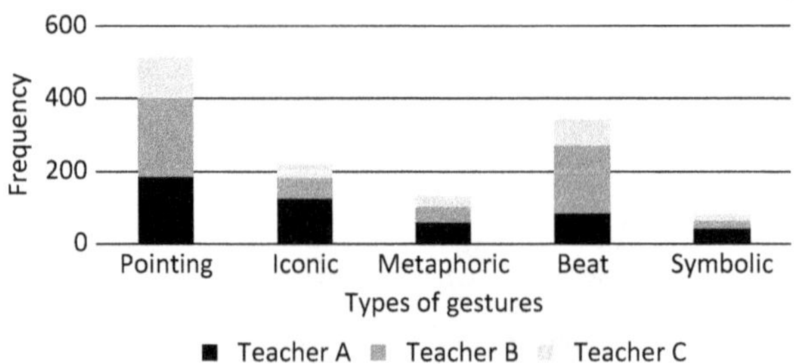

Figure 10.1 Frequency of gestures used for a period of ten observed lessons in the 2018 study.

Table 10.1 Summary of some of the teachers' utterances with respect to the roles of gestures that they used in the 2018 study

Gestural categories	*Role of gesture according to the teachers' perspectives*
Pointing	I point to the learners to give me the answers. I point to what I wrote on the chalkboard ... or to pictures hanging in the class. I also point to the activities on the worksheet or textbook. I point to grab learners' attention. I point to objects outside the classroom such as trees, wind pumps etc. to illustrate what I am explaining.
Iconic	I use gestures to illustrate what I am saying. I use a chair to demonstrate what is above and below. I place my hands under or on top of the chair. Iconic gestures add meaning to what I am saying. I use the bodies of the learners in relation to objects in the class to illustrate concepts such as in front, behind, below, above, on top, underneath, inside, outside, far, near, right etc. Movement of hands to give instructions as to what to do.
Metaphoric	I use these gestures to reinforce what I am saying and affirm my ideas and actions. I often use my hands to demonstrate mathematical relationships such as length (short and tall), volume (small and big). I use bottle tops as a proxy for numbers. I manipulate these bottle tops as metaphors for addition and subtraction. The learners use available material to construct numbers.
Beat	I use these gestures when I am looking for a word that is in my mind ... or to grab learners' attention. I use song and music to reinforce mathematical patterns.
Symbolic	I hold my fingers on my lips to tell learners to keep quiet. I use my hands without talking.

Haipinge's case study (2021) went a step further, in an attempt to interpret and analyse the gestures more deeply. The aim was to identify how they contributed to the mathematical communication of three purposefully selected mathematics teachers. This case study also took place in northern Namibia and was conducted at two schools at a junior secondary level. Data was also generated through a series of five lesson observations per teacher, and stimulus-recall interviews. In order to interpret the gestures beyond just a description of their types, Haipinge (2021) adapted Rosborough's (2010) taxonomy of the functions of gestures and superimposed this on the now famous McNeill (1992) taxonomy. This enabled Haipinge (2021) to interpret each type of gesture in terms of its mathematical function as well. Rosborough's (2010) taxonomy of gesture functions adapted by Haipinge (2021) included the following:

- Joint attention or shared attention (JA): this takes place where one individual communicates with another and vice versa.
- Interpersonal & intrapersonal communication (IC): the former is when a teacher communicates with the learners accompanied by gestures, and the latter is when the teacher communicates and gestures to himself.

- Vocabulary and content development (VCD): these are gestures that reinforce content.
- Transformation activities (TA): these are gestures that connect mathematics to real-life contexts.

As Haipinge's (2021) focus was on analysing how gestures were used to communicate mathematically, perhaps it was not surprising that in his series of lessons the metaphoric gesture was the most prevalent (see Figure 10.2). As his observed classes were in the secondary school grades (where content mathematics is more explicitly foregrounded than in the junior primary grades), this could be another reason why the metaphoric gesture was more prevalent. Related to this, it is thus significant to observe that the VCD gesture (the one that reinforces content) was the most evident – see Figure 10.3.

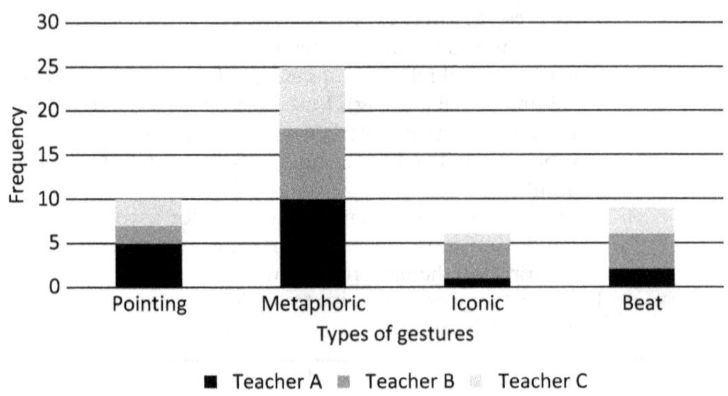

Figure 10.2 Frequency of gestures used for a period of five observed lessons in the 2020 study.

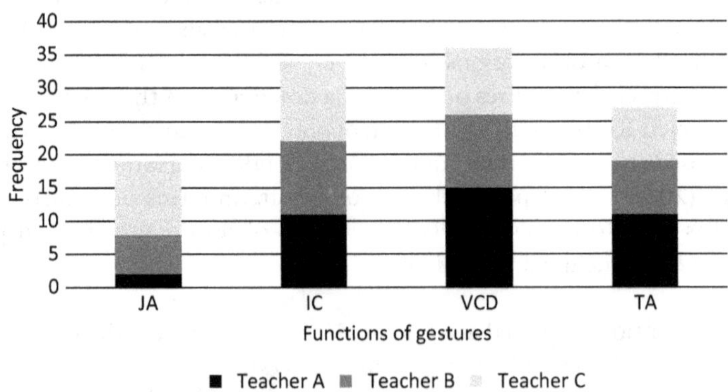

Figure 10.3 Frequency of the functions of gestures used for a period of five observed lessons in the 2020 study.

An example from an Italian Classroom: Gestures for Making a Structure Visual

The second example focuses on students' and teacher's gesturing during a classroom discussion. In particular, we present an excerpt from a discussion in Grade 5, focusing on processes of generalisation in early algebra. The same data was previously used to analyse the roles played by the teacher to foster students' ability to generalise and then produce mathematical arguments. These results can be found in Cusi & Sabena (2020).

Our focus is a discussion that developed after students solved the 'Rivabella parking lots' task (Figure 10.4):

The 'Rivabella parking lots' task required students to study a specific sequence in which trees and parking lots alternated according to the general structure TPPTPPTPPT...TPPT (every sequence starts and ends with a tree), and to justify their conjectures.

All nine groups identified the correct answer to the question – '72 parking lots'. Four out of the nine groups referred to the expression (37–1)×2, and some of them also suggested appropriate arguments. The other four groups referred to the expression 37×2–2. Two of these groups proposed incomplete or partly incorrect lines of reasoning. Lastly, one group proposed 36×2, without giving a logical argument.

After the students tackled the task in their small groups, the teacher-researcher (who took the role of the teacher) selected some answers related to the expression 37×2–2, projected them onto the interactive whiteboard (IWB) and prompted students to identify and explain the other students' reasoning (underlined words were spoken in conjunction with gestures):

1. *T-R: Did those who did not reason in this way, understand why they did 37×2–2?*
2. *S: In my opinion it is right to do times 2 minus 2 because if...there are all the trees that...between two trees there are two parking lots, but between the first and the last...between the first and the last one there are not two parking lots.*
3. *T-R: Would you like to come at the whiteboard, so you explain with the drawing, which may be easier?*
   (S agrees to come at the whiteboard to explain her idea).
4. *T-R: Let's take one of these drawings and use it to reason.*

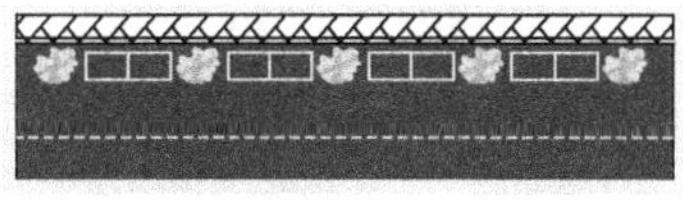

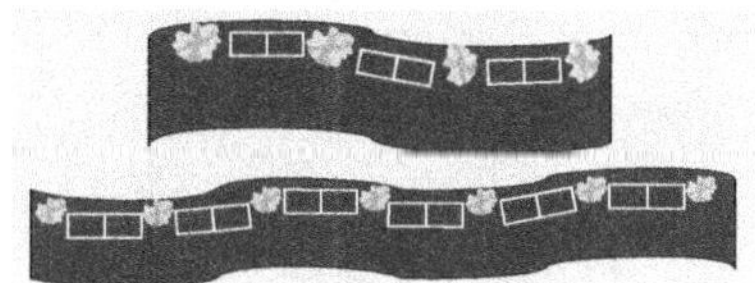

Figure 10.4 The Rivabella Parking lots task.

5. *S: Because…in between these two, for instance (pointing quickly to two consecutive trees in the drawing), there is a … a parking space (touching the space between two trees in the drawing), but between the first (touching the first tree of the sequence) and the last one (touching the last tree of the sequence), here and here (pointing in the same way to the first and to the last tree), there is no parking space, so minus 2 (almost whispering).*
6. *T-R: So, S was saying: "Between the first one and the last one there is no parking space, so I have to take away 2!" (making a gesture using a rapid rotating movement with her right hand).*

The teacher-researcher (T-R) in line 1 poses a question aimed at helping the students interpret the reasoning of their classmates. This was done in order to help the students understand possible effective strategic approaches adopted by their peers in solving the problem. In her answer, S (line 2) does not refer to the particular case of 37 trees, but to a general 'rule': '*times 2, minus 2*'. She is able to grasp the general structure of the formula. However, her reasoning to justify the strategy (in which she explicitly refers to the role played by the first and last trees) is incomplete. Therefore, the T-R asks her to come to the IWB (line 3) and suggests that she refer to the drawings (lines 3–4) in order to make her thinking more visible and share it with the teacher and the classmates.

Then, S also made use of the drawn diagrams and while explaining her line of reasoning, she performed two quick pointing gestures to two consecutive trees 'inside' the drawing. She was considering whether there were any pairs of consecutive trees in the line, except for the first and the last ones. For any such pairs of trees, she identified a space, which she referred to with the words '*parking space*' (line 5). Her pointing gesture physically touching the drawing precisely on the line separating the two consecutive parking spaces indicates that she actually meant the space containing two parking spaces. Through her speech and gestures, and by referring to the given figures, she portrayed the relationship between two consecutive trees and one space between them. Then, she identified the first and the last trees as somehow different. Even when S was physically referring to a specific drawing (by means of her gestures), she was actually thinking more generally.

The teacher then aimed at focusing the students' attention on S's argument and reflecting on its accuracy and completion. By repeating the same words and by using similar gestures, the teacher emphasised only the part of S's argument which needed clarification. Her final gesture (rotating) is not referred to in the figure. This possibly signals that something needed to be clarified in the corresponding part of the argument.

After another student had tried to interpret the diagram and explain his reasoning, he then asked for more time to reflect. M asked to intervene:

15. *M: Well, let's take the same example. Between here and here (quickly pointing to the first and to the second tree) there is a parking space. We…*
16. *T-R (interrupts): That is, two parking spaces, you mean. A space with two parking spaces.*

17. *M: Yes. There are seven (counting the trees).*
18. *T-R: Seven trees.*
19. *M: Seven trees. But here (pointing to the last tree of the sequence) there are no more parking spaces (pointing to the empty space at the right of the last tree in the sequence).*
20. *T-R: You say that to the right of the last tree (pointing to the last tree) there are no parking spaces.*
21. *M: Yes.*
22. *T-R: And so?*
23. *M: We must stop and take two away: there is [no parking space] at the beginning, nor at the end.*
24. *T-R: Tell me if I've understood correctly: M says, "To the right of every tree there are two parking lots (placing left forefinger and right index and thumb on the drawing, as shown in Figure 10.5)". But if I have seven trees and I multiply by two (keeping the left hand still and moving the right hand to the right), I am also counting the spaces that are here (gestures as in Figure 10.6), that must be taken away. Did M reason correctly?*

(The gestures in Figures 10.5 and 10.6 are not the original ones recorded in the classroom and have been recreated to improve their resolution.)

Most of the students declared that M was correct, so the teacher asked some of them to once again suggest a reason for, in particular, the role played by '–2'

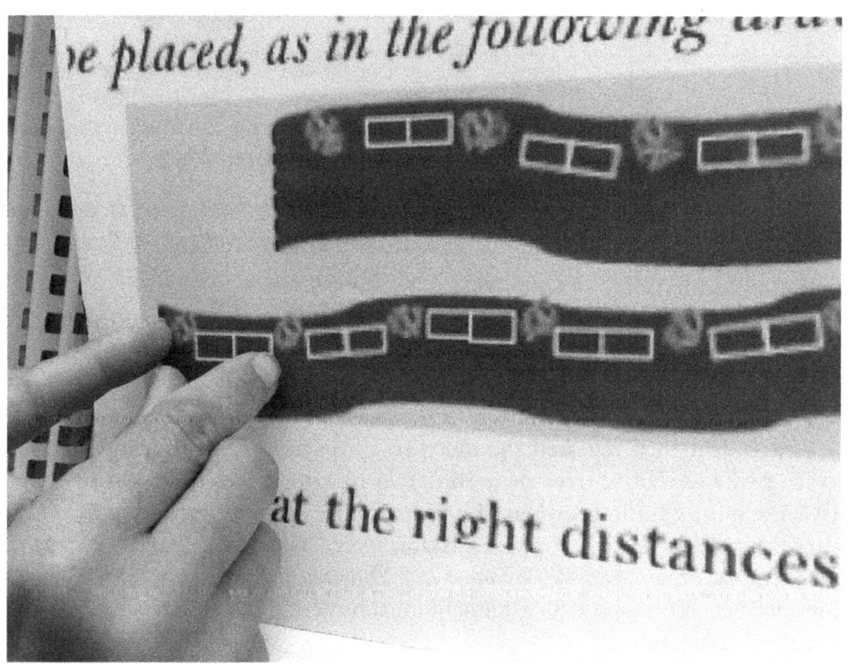

Figure 10.5 S indicates the position of the trees relative to the parking spaces.

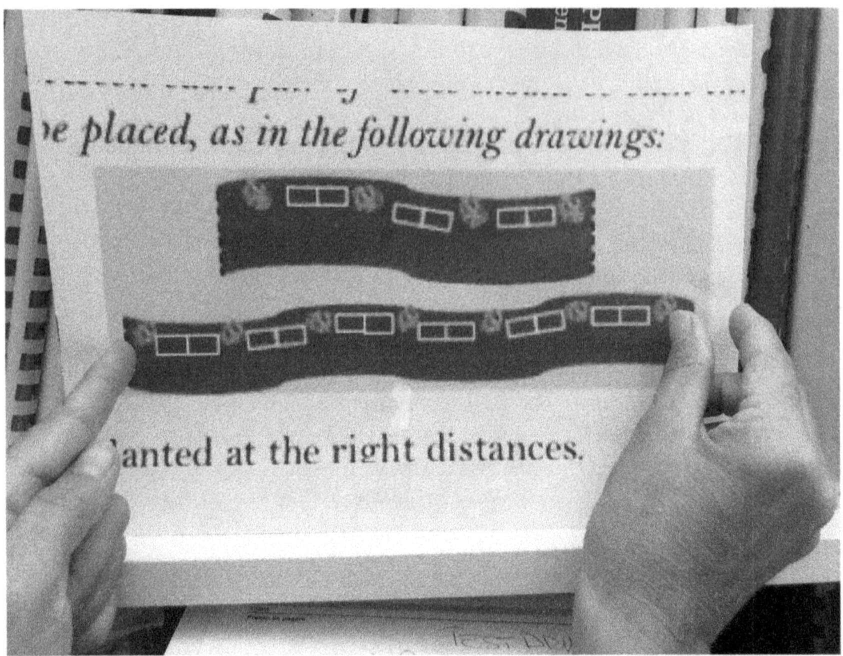

Figure 10.6 S indicates the number of parking spaces and trees by moving her right hand to the right.

in the expression 37×2–2. After some suggestions from his classmates, she asked M to explain more clearly why 2 should be subtracted from 37×2:

25. *M: I should always do times two, but also the last one is times two, so I am also counting a …parking space (pointing to the right of the figure) that does not exist.*

M began by saying that he was referring to the same example as S (line 15), but actually chose another one. This is another indication that, prompted by the specific focus on explaining the interpretation of the given formula with respect to the figures, the students took advantage of the assigned figures to reason generally.

The student then repeated the first part of S's argument, showing that between two consecutive trees there were two parking spaces (lines 15–18). Line 19 is the most significant, when M pointed to the last tree and emphasised that after this tree there were no more parking spaces This explains the '–2' part of the formula, or, as M said, "*We must stop and take 2 away*" (line 23). Then M continued to also propose S's argument for subtracting two, referring to the first and to the last tree.

With the aim of making students focus on M's argument and reflect on it, in lines 16 and 18, T-R reformulated parts of M's argument to make it clearer.

This is evident when, by means of a semiotic game, T-R (line 20) interpreted M's gestures, adding a clue – "*to the right*". Her aim was to support him in his reasoning, which was based on considering two parking lots to the right of every tree, except for the last one (line 23).

By means of an interplay between words, gestures and references to the diagrams, the T-R made the given formula correct, clear and complete for all the students. First, she repeated only the (correct) part of M's argument (line 24), introducing the word "*every*". She used gestures too, in synergy with her words: in Figure 10.5 we can see how she purposely coordinated her two hands in order to stress the relation between one tree (pointed at with her left hand) and two parking lots (indicated with the fingers of her right hand), which provides a narrative for the '7×2' part of the formula. This is also emphasised in the subsequent gesture, in which T-R's right hand is motioning towards the right: this can be interpreted with reference to the mathematical operation 'times 2', through a possible metaphoric reference to the number line. At the same time, it keeps an iconic link with the way the diagram is being looked at (for any tree, considering the space at its right). Her left hand, pointing and motionless, provides a reference to the diagram. With the same emphasis, the T-R repeated the same right-handed gesture after the last tree (Figure 10.6), while she spoke more slowly. This repetition of gestures, or 'catchment' (McNeill et al., 2001), is used by the T-R to highlight an algebraic structure in the discourse to model the situation.

As a result, M also felt the need to correct his previous argument (line 25), that if 37 was multiplied by 2, two more 'imaginary' parking lots would be added to the right of the last tree.

Conclusion

From our discussion and deliberation in the above lesson vignettes of three teachers from different parts of the world, what can we gain with respect to mathematics teaching-learning processes in the classroom? It is evident that gestures are integral to who the teacher is, and what and how the teacher teaches. The vignettes show that the medium of gestures is not only an intricate and complex process of teaching, but part and parcel of a teacher's identity. The notion that gestures form an important and vital medium for communication, teaching and learning is gaining traction in research circles in mathematics education. This is encouraging as we argue that gestures are integral to the Semiotic Bundle and may contribute, in synergy with other semiotic resources, to mathematics reasoning and communication.

We argue that research has not yet interrogated in sufficient depth the question of how the spontaneous use of gestures relates to a purposely directed use of them. This question is particularly important in the teaching context. The semiotic game (Arzarello et al., 2009) may be one way to integrate gestures when working with students to enhance mathematics communication and learning. Recognising that these bodily actions are manifestations of the entire body's role in generating and understanding knowledge should prompt teachers to overcome the hegemony of speech and symbols and broaden their focus of attention in the classroom. Students' gestures should be included as part and parcel of their mathematical learning and activities. As Alibali Alibali and Nathan (2012, p. 247) so appropriately observe, "gestures are taken as evidence that the knowledge itself is embodied and the body is involved in thinking". Communication is key, not only in mediating knowledge, but also in making knowledge accessible to all. Therefore the recognition of gestures as a legitimate vehicle for transmitting and understanding knowledge will go a long way towards ensuring epistemological access to learners in situations where the efficacy of the spoken language is compromised, for whatever reason. Even if research on the contribution that gestures may have on bi- or multilingual students' mathematics learning is still emerging and growing, we agree with Robutti and colleagues in recognising a potential in this direction.

We would like to thank Annalisa Cusi for giving her consent to publish part of the data we gathered together and for allowing us to reveal her face in the excerpts of selected video clips. We also would like to thank Dietlinde Namakalu and David Haipinge whose data was also used in this chapter.

References

Alibali, M. W., Kita, S., & Young, A. J. (2000). Gesture and the process of speech production: We think, therefore we gesture. *Language and Cognitive Processes, 15*, 593–613. https://doi.org/10.1080/016909600750040571

Alibali, M. W. (2005). Gesture in spatial cognition: Expressing, communicating, and thinking about spatial information. *Spatial Cognition and Computation, 5*(4), 307–331. https://doi.org/10.1207/s15427633scc0504_2

Alibali, M. W., & Nathan, M. J. (2012). Embodiment in Mathematics teaching and learning: Evidence from learners' and teachers' gestures. *Journal of the Learning Sciences, 21*(2), 247–286. https://doi.org/10.1080/10508406.2010.610446

Antle, A. N., Corness, G., & Droumeva, M. (2008). What the body knows: Exploring the benefits of embodied metaphors in hybrid physical digital environments. *Interacting with Computers, 21*(1, 2), 66–75. https://doi.org/10.1016/j.intcom.2008.10.005

Arcavi, A. (2003). The role of visual representations in the learning of mathematics. *Educational Studies in Mathematics, 52*(3), 215–241. https://doi.org/10.1023/A:1024312321077

Arzarello, F. (2006). Semiosis as a multimodal process, *Revista latino americana de investigación en matemática educativa, Vol. Especial*, 267–299.

Arzarello, F., Paola, D., Robutti, O., & Sabena, C. (2009). Gestures as semiotic resources in the mathematics classroom. *Educational Studies in Mathematics, 70*, 97–109. https://doi.org/10.1007/s10649-008-9163-z

Castellon, V., & Enyedy, N. (2006). *Teach and Algebra Classroom.* San Francisco, CA: American Educational Research Association.

Chikiwa, C., & Schäfer, M. (2020). The use of gestures and language as co-existing visualisation teaching tools in multilingual. *The 14th International Congress on Mathematical Education, Shanghai, 12–19 July, 2020.*

Cusi, A., & Sabena, C. (2020). The role of the teacher in fostering students' evolution across different layers of generalisation by means of argumentation. *RECME-Revista*

Colombiana de Matemática Educativa, 5(2), 93–105. ISSN 2500-5251, http://funes.uniandes.edu.co/22718

Di Paolo, E., Rohde, M., & De Jaegher, H. (2007). Horizons for the enactive mind: Values, social interaction and play. In: J. Stewart, O. Gapenne, & E. Di Paolo (Eds.) *Enaction:Towards a New Paradigm for Cognitive Science* (pp. 33–87). Cambridge, MA: MIT Press.

Duval, R. (2006). A cognitive analysis of problems of comprehension in a learning of mathematics, *Educational Studies in Mathematics, (61)*, 103–131. https://doi.org/10.1007/s10649-006-0400-z

Edwards, L. (2009). Gestures and conceptual integration in mathematical talk. *Educational Studies in Mathematics, 70*(2), 127–141. https://doi.org/10.1007/s10649-008-9124-6

Ernest, P. (2006). A semiotic perspective of mathematical activity. *Educational Studies in Mathematics, 61*, 67–101.

Freudenthal, H. (1991). *Revisiting Mathematics Education: China Lectures.* Dordrecht: Kluwer.

Gallese, V., & Goldman, A. (1998). Mirror neurons and the simulation theory of mind-reading. *Trends in Cognitive Sciences, 12*, 493–501. https://doi.org.10.1016/s1364-6613(98)01262-5

Garber, P., & Goldin-Meadow, S. (2002). Gesture offers insight into problem-solving in adults and children. *Cognitive Science, 26*, 817–831. https://doi.org/10.1016/S0364-0213(02)00087-3

Goldin-Meadow, S. (2000). Beyond words: The importance of gesture to researchers and learners. *Child Development, 71*, 231–239. https://doi.org/10.1010/1467-8624.00138

Goldin-Meadow, S. (2003). *Hearing Gesture: How Our Hands Help Us Think.* Cambridge, MA: Belknap Press.

Goldin-Meadow, S. (2006). Talking and thinking with our hands. *Current Directions in Psychological Science, 15*(1), 34–39. https://doi.org/10.1010%2Fj.0963-7214.2006.00402.x

Goldin-Meadow, S. (2015). From action to abstraction: Gesture as a mechanism of change. *Developmental Review, 38*, 167–184. https://doi.org/10.1016/j.dr.2015.07.007

Goldin-Meadow, S., Kim, S., & Singer, M. (1999). What the teacher's hands tell the student's mind about math. *Journal of Educational Psychology, 91*, 720–730. https://doi.org/10.1037/0022-0663.91.4.720

Haipinge, D. T. (2021). *An analysis of selected Grade 8 mathematics teachers' use of gestures as visualisation tools to support mathematical meaning-making.* Unpublished Master's thesis, Rhodes University: Grahamstown.

Iverson, J. M., & Goldin-Meadow, S. (2001). The resilience of gesture in talk: Gesture in blind speakers and listeners. *Developmental Science, 4*(4), 416e422.

Kendon, A. (1980). Gesticulation and speech: Two aspects of the process of utterance. In: M. R. Key (Ed.) *The Relation Between Verbal and Nonverbal Communication* (pp. 207–227). The Hague: Mouton.

Kendon, A. (1988). How gestures can become like words. In: F. Poyatos (Ed.) *Cross-Cultural Perspectives in Nonverbal Communication* (pp. 131–141). Toronto: Hogrefe.

Kita, S. (2000). How representational gestures help speaking. In: D. McNeill (Ed.) *Language and Gesture: Window Into Thought and Action* (pp. 162–185). Cambridge, UK: Cambridge University Press

Kress, G. (2004). Reading images: Multimodality, representation and new media. *Information Design Journal, 12*(2), 100–109. https://doi.org/10.1075/idjdd.12.2.03kre

Lakoff, G., & Nùñez, R. (2000). *Where Mathematics Comes From: How the Embodied Mind Brings Mathematics into Being.* New York: Basic Books.

McNeill, D. (1992). *Hand and Mind: What Gestures Reveal About Thought.* Chicago: University of Chicago Press.

McNeill, D. (2005). *Gesture and Thought.* Chicago: The University of Chicago Press.

McNeill, D., Quek, F., McCullough, K. E., Duncan, S., Furuyama, N., Bryll, R., Ma, X. F., & Ansari, R. (2001). Catchments, prosody and discourse. *Gesture, 1*(1), 9–33. https://doi.org/10.1075/gest.1.1

Namakalu, D. N. (2018). *An analysis of the roles and functions of teachers' gestures as visualisation tools in the teaching of mathematics at the junior primary phase (Grades 0 –3).* Unpublished master's thesis, Rhodes University: Grahamstown.

Nemirovsky, R., & Noble, T. (1997). On mathematical visualization and the place where we live. *Educational Studies in Mathematics, 33*(2), 99–131. http://dx.doi.org/10.1023/A:1002983213048

Otte, M. (2006). Mathematical epistemology from a peircean semiotic point of view. *Educational Studies in Mathematics, 61*, 11–38.

Özçaliskan, S., & Goldin-Meadow, S., (2009). When gesture-speech combinations do and do not index linguistic change. *Language and Cognitive Processes, 28*(24), 190–217. https://dx.doi.org/10.1080%2F01690960801956910

Peirce, C. S. (1958). *Collected Papers, Vol. I-VIII* (p. 2). Cambridge, MA: Harvard University Press.

Radford, L. (2009). Why do gestures matter? Sensuous cognition and the palpability of mathematical meanings. *Educational Studies in Mathematics, 70*(2), 101–126. https://doi.org.10.1007/s10649-008-9127-3

Radford, L., Edwards, L., & Arzarello, F. (2009). Beyond words. *Educational Studies in Mathematics, 70*(3), 91–95. https://doi.org.10.1007/s10649-008-9172-y.

Rizzolatti, G., & Gallese, V. (1997). From action to meaning. In: J.-L. Petit (Ed.) *Les neurosciences et la philosophie de l'action* (pp. 217–229). Paris: J. Vrin.

Robutti, O., Sabena, C., Krause, C., Soldano, C., & Arzarello, F. (accepted). Gestures in mathematics thinking and learning. In: M. Danesi (Ed.) *Handbook of Cognitive Mathematics.* Cham: Springer.

Rosborough, A. (2010). Gesture as an act of meaning-making: An eco-social perspective of a sheltered-English second grade classroom. *ProQuest Dissertations and Theses, 282.* http://dx.doi.org/10.34917/2023344

Roth, W. M. (Ed.). (2009). *Mathematical Representation at the Interface of Body and Culture.* Charlotte, NC: Information Age Publishing.

Sabena, C., Robutti, O., Ferrara, F., & Arzarello, F. (2012). The development of a semiotic frame to analyse teaching and learning processes: examples in pre- and post-algebraic contexts. In: L. Coulange, J-P. Drouhard, J-L. Dorier, & A. Robert (Eds.) *Recherches en Didactique des Mathématiques, Numéro spécial hors-série, Enseignement de l'algèbre élémentaire: bilan et perspectives* (pp. 231–245). Grenoble: La Pensée Sauvage.

Schäfer, M., & Namakalu, N. (2019). The role of gestures in the teaching of mathematics. *Proceedings of 53 Jahrestagung der Gesellschaft für Didaktik der Mathematik (GDM).* Regensburg: Germany.

Tall, D. (2000). Biological brain, mathematical mind & computational computers (how the computer can support mathematical thinking and learning). In: W-C. Yang, S-C. Chu, & J-C. Chuan (Eds.) *Proceedings of the Fifth Asian Technology Conference in Mathematics*, Chiang Mai, Thailand (pp. 3–20). ATCM Inc., Blackwood VA.

Varela, F. J., Thompson, E., & Rosch, E (1991). *The Embodied Mind: Cognitive Science and Human Experience.* Cambridge, MA: MIT Press.

Volterra, V., Caselli, M.C., Capirci, O., & Pizzutto, E. (2005). Gesture and the emergence and development of language. In: M. Tomasello, & D. Slobin (Eds.) *Beyond Nature-Nurture. Essays in Honor of Elizabeth Bates* (pp. 3–40). Mahwah, NJ.: Lawrence Erlbaum Associates.

Vygotsky, L. S. (1931/1978). *Mind in Society. The Development of Higher Psychological Processes.* In: M. Cole, V. John-Steiner, S. Scribner, & E. Souberman (Eds.) Cambridge, MA, and London: Harvard University Press.

Wilson, M. (2002). Six views of embodied cognition. *Psychonomic Bulletin & Review, 9*(4), 625–636. https://doi.org/10.3758/BF03196322

11 Gestures, Diagrams and Verbal Language in Multilingual Classrooms

Clemence Chikiwa

Introduction

This chapter explores how one teacher of a multilingual class used verbal language, gestures and diagrams to communicate and enhance epistemological access to mathematical concepts during teaching. In teaching and learning, especially in rural and township multilingual mathematics classes, communication is done through a multimodal system of signs and symbols to provide an instructionally inclusive environment. The term 'multimodal' in this chapter is used with reference to multiple modalities that are employed to communicate and express meanings to our interlocutors, such as words, sounds, images and so on (Kress, 2004). As explained by Radford et al. (2009),

> these resources or modalities include both oral and written symbolic communication as well as drawings, gestures, the manipulation of physical and electronic artefacts, and various kinds of bodily motion.
>
> (pp. 91, 92)

A multimodal perspective on mathematics teaching and learning is addressed through what Arzarello et al. (2009) call the 'semiotic bundle lens'. A semiotic bundle is a system of signs produced by one or more interacting individuals (Arzarello, 2006). In school settings, a semiotic bundle is made of the signs that are produced by teachers and their learners while interacting during the teaching and learning of mathematics.

It is important to pay attention to a wide variety of means of expression, from the mathematical diagrams and symbols to the embodied ones like gestures and gazes, and consider them as semiotic resources in teaching and learning processes. This provides access to mathematical knowledge, especially in multilingual classes. Most rural and township schools in Southern Africa consist of learners who are not learning in their first language. Mathematics is, in most cases, taught in the language of learning and teaching (LOLT), which is neither the teachers' nor the learners' first language and one in which learners lack proficiency. The use of an LOLT in which learners are not proficient has created inequalities at various fronts, resulting in various forms of learning difficulties

DOI: 10.4324/9781003172420-15

associated with access to mathematical concepts. In many cases teachers of such multilingual classes have found ways to help their learners understand the intended content. Some of the ways include code switching, code mixing, code meshing and translanguaging, among others.

Mathematics is a discipline that is interested in inventing and using various signs such as gestures, symbols and diagrams, especially in the teaching and learning process. These gestures, diagrams and signs are employed to enhance understanding of mathematical concepts during interaction. Mathematics and its teaching and learning relies on an intensive use of different kinds of signs and symbols such as letters and cyphers for numbers, gestures, diagrams and formulae. Such signs and symbols can also be in the form of gestures produced by teachers and learners during instruction. Teaching and learning mathematics without using any language, gestures, signs, symbols, diagrams or handling manipulatives does not seem to be possible at all.

In their book, *The Culture of Diagram*, Bender and Marrinan (2010) investigate the interplay between words, pictures and formulae in teaching and learning mathematics. They conclude that diagrams, together with correct verbal language, appear to be valuable tools to understand this crucial interplay. In this same book, Bender and Marrinan (2010) show in detail the role that diagrams play as means to construct knowledge and interpret data and equations in mathematics. The use of diagrams, together with gestures and other forms of visualisation processes, are even more important in the majority of Southern African mathematics classes where learners are not proficient in the LOLT.

The relation between the use of verbal language, together with visual strategies such as gestures and other artefacts and learners' mathematical understanding and reasoning, may open avenues for thinking about improving access in mathematics education. This chapter focuses on the use of verbal language together with visual strategies, to provide learners with access to mathematical knowledge in high school. While learners have access to schooling as enshrined in the constitutions of most Southern African countries, the most crucial and needed element is epistemological access for success in mathematics at the school level. This is particularly pertinent when learners are enabled to become successful participants in an academic practice (Morrow, 2007). Many studies recognise the importance of the role that visualisation in its various forms (including gestures and diagrams) plays in heightening conceptual understanding, problem solving and mathematical reasoning in multilingual mathematics teaching environments.

Visualisation in Mathematics Teaching

Visualisation is regarded as an inherent and integral component of teaching and learning. Gutiérrez (1996) defines visualisation as the set of types of images, processes and skills necessary for learners of mathematics to produce, analyse, transform and communicate visual information related to objects, models and mathematical concepts. This set of images can be in the form of diagrams and/or gestures used during mathematics teaching. Arcavi (2003) provides a more comprehensive definition of visualisation as:

> the ability, the process and the product of creation, interpretation, use of and reflection upon pictures, images, diagrams, in our minds, on paper, or with technological tools, with the purpose of depicting and communicating information, thinking about and developing previously unknown ideas and advancing understandings.
>
> (p. 217)

The use of pictures, images and diagrams is highlighted in these definitions. Within these descriptions, communication is crucial when representations are chosen, and as Nardi (2014) notes, the process of visualisation can be external or internal. Visual processes may be internal (in the mind) or external on paper, chalkboard or with technological tools. Of interest in this chapter is how external visualisation (gestures and diagrams) interplays with verbal language in multilingual classes. These external visualisations may be a display of static or virtual visual artefacts such as physical manipulatives, drawings, charts, diagrams and animated manipulatives which are commonly used and are important in enhancing understanding (Cohen & Hegarty, 2007).

Gutiérrez (1996, p. 10) identifies four key elements of visualisation. These are, first, mental images which are "any kind of cognitive representation of a mathematical concept or property by means of visual or spatial elements"; second, external representations which are "any kind of verbal or graphical representation of concepts or properties including pictures, drawings, diagrams, etc. that helps to create or transform mental images and to do visual reasoning"; third, the process of visualisation and this is "a mental or physical action where mental images are involved"; and, last, the visualisation abilities of visualisation. These are stable capacities of mathematics which are necessary for effective learning of the subject. For learning to occur, a smooth interaction between these elements is required. Visual processes are enhanced by the presence of correct images, pictures, gestures and diagrams as external representations.

Research has consistently shown how mathematicians owe their success to visualisation (Arcavi, 2003; Gutiérrez, 2018). During teaching, various visuals are used, and learners are encouraged to visualise during the process by using diagrams, graphs and gestures to discover mathematical patterns and relationships (Tall, 2004). Visualisation is thus useful in understanding and learning many other content areas in school mathematics (arithmetic, algebra, functions, statistics, etc.). This is because all school mathematics domains may benefit from the use of visual representations such as graphs, diagrams, drawings and dynamic representations of calculations (Gutiérrez, 2018). Visualisation plays a pertinent role in making the concepts accessible to learners.

It is important that learners be exposed to correct mental images that will help them associate those images with their corresponding verbal descriptions. Gutiérrez (1996) asserts that a "mental image is a mental representation of a mathematical concept or property containing information based upon pictorial, graphical or diagrammatic elements" (p. 6). Thus, if incorrect representations in the form of gestures and diagrams are used, mathematical concepts will thus be incorrectly represented in their minds and hence externally by learners. This is even more crucial in multilingual settings where most learners do not learn mathematics in their first language.

Gestures and Multilingual Mathematics Classroom Teaching

Gestures are crucial as they play the role of dynamic diagrams: therefore they can be regarded as visual in nature. As alluded to by Sinclair and Tabaghi (2010), gestures can be used to sketch objects and their relations during teaching. McNeill (1992) found speech and gestures to be closely linked in many respects and that gesture and speech proceed together during interactions. Gestures have many functions, especially in the teaching and learning environment. According to McNeill (1992), "gestures do not just reflect thought but have an impact on thought. Gestures, together with language, help constitute thought" (p. 242). Studies in multilingual settings have shown that in situations where the LOLT is the second or third language of the learners, teachers often use the first language of these learners as a resource to aid understanding and provide access to mathematical concepts. This is frequently done through teacher code switching accompanied by gestures. In this chapter, code switching is the alternate use of more than one language in the same conversation or utterance (Adler, 2001).

In multilingual classes, especially in Southern Africa, teachers have been observed using their learners' first language for promoting meaningful teaching. This is because in most rural and township schools, teaching occurs in a language which is neither the teachers' nor the learners' mother tongue. It has been noted that the LOLT chosen by schools is often a language which learners are still learning and are consequently not proficient in. This has resulted in teachers devising ways in which they can help their learners improve their understanding of mathematics under such constraints. Teachers thus use their learners' first language together with other visual strategies such as gestures and diagrams.

McNeill (1992) speculated that "gestures are an integral part of language as much as are words, phrases and sentences" (p. 2). My focus on teachers' gestures and diagrams as visualisation tools in this chapter is informed by the large body of research showing that gestures, when properly used with other semiotic tools, contribute to communication in a wide range of settings (Kendon, 2004; Sabena, 2018). The use of visualisation processes such as gestures together with diagrams in teaching mathematics is an important component to help learners to build their own adequate internal representations of mathematical ideas. Gestures help to make visible the mental pictures of concepts. I suggest that such use of gestures is particularly relevant in multilingual classes to aid understanding and access to mathematical ideas where the LOLT is not well developed in learners.

Several studies have pointed towards the importance of gestures' ability to communicate concepts and the visual appearance of mathematical objects, such as diagrams, shapes and slopes on the line graphs (Sabena, 2018; Yoon et al., 2011). Gestures are crucial for communicating and constructing mathematical understanding (Yoon et al., 2011) especially in situations where learners are not proficient in the LOLT. Thus, gestures, in addition to drawings and pictures, have the potential to play a mediating role in the visualisation of ideas and concepts when teachers explain concepts to learners in multilingual settings (Kendon, 2004). This, in turn, is envisaged as increasing access to learning and providing an equitable ground for learning mathematics in a language other than their first language. Work by Alibali et al. (2013) and that by Hare and Sinclair (2015) suggest that a combination of speech, gestures and diagrams provides real-time evidence of the meanings that gesturing and diagramming help to create in mathematics.

Since gestures, together with verbal language and other visuals, help constitute thought (McNeill, 1992), they may reflect both what the teacher' thoughts about his/her learners at a particular moment are and what the learners themselves may be struggling with at that same moment. Goldin-Meadow and Alibali (2013) raise another important point that "gesture can change speakers' thoughts" (p. 247). When teachers use gestures appropriately in teaching mathematics, the process has the potential to change the teacher's thinking about the concepts he/she is referring to, and this may result in this same speaker altering or revising the course of thinking and learning. Such refinements are vital for making conceptual learning possible during teaching. Gestures, diagrams and speech are thus all combined in the construction of complex multi-layered and multimodal representations in which no single layer is complete or coherent on its own.

In this chapter, I concur with the notion that teachers' concurrent use of gestures, diagrams and verbal communication in both the LOLT and learners' first language provides one way of providing learners with access to meaningful mathematics learning. Such use of various language repertoires is envisaged as one way in which social justice and equity in mathematics teaching can be realised. It is therefore important that the appropriate use of gestures together with the learners' first language be recognised as a legitimate teaching strategy that supports good teaching, and I argue that in order for gestures and verbal language to be meaningful, they should be used strategically (Chikiwa & Schäfer, 2019). Studying the concurrent use of verbal language and gestures in multilingual classes is crucial because gestures can serve as important support mechanisms to teachers who teach in a language that is not their first nor the learners' first, and who thus have a low proficiency in the LOLT. Gestures can also provide supporting backup for teachers who find themselves in situations where they lack proficiency in the first language of their learners and are not familiar with the associated syntax, pronunciation and intonation practices and routines of that language. Thus, all endeavours to promote multilingual class teaching that will provide meaningful access into mathematical concepts needs to be explored and encouraged.

Classification of Gestures

Although gestures have been classified in many ways in the research literature (see Ekman & Friesen, 1974; Scherer & Ekman, 1982; Lim, 2019; Kendon, 2004; Cienki & Müller, 2008; Goldin-Meadow, 2015), in this chapter I have adopted McNeill's (1992) classification of four types of hand gestures: *iconic, metaphoric, deictic* and *beat* gestures.

1 *Iconic* gestures are hand movements used to create a picture or present images of concrete entities and/or actions. McNeill (2005) elaborates further, referring to iconic gestures as hand gestures that represent meaning closely related to the semantic content sections of the speech that they accompany. Iconic gestures draw their communicative power from being perceptually similar to the event or concept under discussion (McNeill, 1992). Iconic gestures can be subcategorised into the following as found in literature:

   - *Action gestures* are gestures that occur when the arms or hands mimic an observable physical act performed by an organism (Edwards, 2009; Arzarello et al., 2009; McNeill, 2005). An example is the movement of the hands when showing the wave motion of water.
   - *Magnitude gestures* occur when the hands are used to represent a large or small distance, size or amount of some phenomenon (McNeill, 1992).
   - *Movement gestures* are those produced when the hands are used to represent an observable entity or entities going from one point to another, with a relative emphasis on direction, speed or spatial position of the entity's (entities') starting point, relative to its (their) ending point (Kendon, 2004) (e.g. using a hand to represent the revolution of the earth around the sun).
   - *Shape gestures* are hand gestures that indicate the physical outline of a form or figure (e.g. a circle) (Kendon, 2004; McNeill, 1992).
   - *Spatial position gestures* are hand gestures that one may use to indicate an observable entity placed in a specific location relative to another observable entity (Gibson, 2014; Arzarello et al., 2009).

   In another study, Edwards (2009) extended McNeill's iconic classification to include iconic-physical gestures (those gestures resembling physical phenomenon) and iconic-symbolic gestures (those referring to mathematical symbols). Arzarello and Robutti (2004) used iconic-symbolic gestures as well, in a similar way to Edwards (2009), but they further extended the categorisation to include iconic-representational gestures which are gestures related to graphic representations of concepts in mathematics.
2 *Metaphoric* gestures indicate a pictorial representation of an abstract mathematical idea that cannot be represented physically.
3 *Deictic (pointing)* gestures are those used to point to or refer to objects, locations, inscriptions or people, with fingers or hands, directing the listener's attention to these objects. Cochet and Vauclair (2014, p. 279) state that "deictic gestures are produced to direct a recipient's attention towards

a specific referent in the proximal or distal environment". Deictic gestures are context dependent, pointing at a concrete item, indicating directions, referring to the past or an abstract loci in space (Alibali et al., 2013).

4 *Beat* gestures are short, rapid and repeated hand movements that follow the rhythm or beat of the speech. These types of gestures may be considered as manual gestures which do not have a clear referential component. They are used mainly to enrich conversations.

The classification given by McNeill (1992) is not based only on the physical features of a given gesture, but by considering the relationships with contextual information. Thus, when looking at any given gesture performed, the interpretation of it should consider the broader context in which a gesture is performed. Therefore the verbal language and the diagrams accompanying the gesture are considered crucial in this chapter. While this classification by McNeill formed the basis of my analysis, other gestures emerged from the actions of the teachers I observed teaching.

Diagrams in Mathematics Teaching

In the teaching of mathematics, diagrams are part and parcel of the process of enhancing understanding. Mathematical concepts are in many cases represented visually in diagram form and hence diagrams are widely used. This implies that diagrams are generally taken to be an integral component of doing and understanding mathematics. Diezmann and English (2001) define a diagram as "a visual representation that displays information in a spatial layout" (p. 77). According to Winn (1987, p. 157), a "diagram is defined as an abstract visual representation that exploits spatial layout in a meaningful way, enabling complex processes and structures to be represented holistically". The use of this concept is by no means restricted to 'images' or 'graphical representations'.

Learners are encouraged to use diagrams during learning for various reasons. This is because diagrams can help unpack the structure of a problem. Diagrams are often credited with helping learners succeed in mathematical problem solving by enabling them to discover and examine underlying relationships (Pantziara et al., 2009) and generate new ideas (Diezmann & English, 2001; Nunokawa, 2006). One advantage of diagrams is their ability to reduce learners' cognitive load (Gibson, 2014; Koedinger et al., 2008). Diagrams convey more information in less time. Visual information helps overcome language barriers (Cohen & Hegarty, 2007) and this is especially important in this study where the environment is multilingual and where learners are synchronously learning mathematical concepts in the LOLT. The use of diagrams, images or pictures to explain mathematical concepts may help to improve understanding and access to concepts for learners learning in a language they are not proficient in.

In mathematics and its teaching, diagrams come in many different forms. Static mathematical diagrams appear on paper, in books, magazines and journals, while dynamic and interactive diagrams make use of the capabilities of digital technologies (Yerushalmy & Naftaliev, 2011). How they are used to enhance learning, especially in multilingual settings, is critically important. Diagrams as external representations are used to illustrate multiple relationships and properties involved in a given problem, at times without suggesting solution procedures (Kadunz, 2006). This is a descriptive role of a diagram in mathematics teaching. In this case the sole function of the diagram is to give a "synoptical apprehension of the properties mentioned in a given problem" (Mesquita, 1998, p. 191). In other cases, the external representation itself acts as a support for intuition, suggesting transformations that lead to a solution. In this case, one may say that the external representation has a heuristic role (Mesquita, 1998). Diagrams are powerful opportunities for grappling with and learning abstract relationships, for example, in one-to-many functions, or in sketching graphs of a family of functions. Mudaly (2012) refers to diagrams as self-explanatory tools for teaching and learning in mathematics. As noted by Carter (2018), diagrams function as objects or signs that can be manipulated and subsequently experimented on, and certain properties and relations in mathematics teaching and learning are directly observable in diagrams. This makes diagrams useful, particularly in multilingual classes where learners are being taught in a language not their first.

A Multimodal Approach to Teaching

This chapter focuses on communication in mathematics multilingual classes. This is usually through a multimodal approach to communication. Teachers use verbal language, gestures and diagrams in the teaching process. Sáenz-Ludlow and Kadunz (2016, p. 14) put forward that:

> ... communication is essentially message exchange that depends, among other things, on: socio-cultural contexts; the content of the message; the language used to convey the message (syntax, grammar and semantics, active and passive lexicon); the means of human interaction (voice-intonation, diagrams and graphs, writing and inscriptions); and the visual means of their delivery (gestures, pointing/deixis, gazing, posture, and the like).

This definition of communication includes the key aspects that this chapter focuses on – that is, verbal communication and the use of gestures and other visuals such as diagrams during the teaching process.

This chapter is informed by the theory of multiple representations proposed by Richard Lesh, referred to as the Lesh Multiple Representations Translations Model (LMRTM). According to Lesh et al. (1987), representations are crucial for understanding mathematical concepts. In their work, Lesh et al. (1987) considered both internal and external representations as important elements for accessing and understanding mathematical concepts. In multiple-representations-based mathematics instruction, manipulatives, real-world situations, oral explanations and static visuals should be used rather than only mathematical symbols. They defined representation as "external (and therefore observable) embodiments of

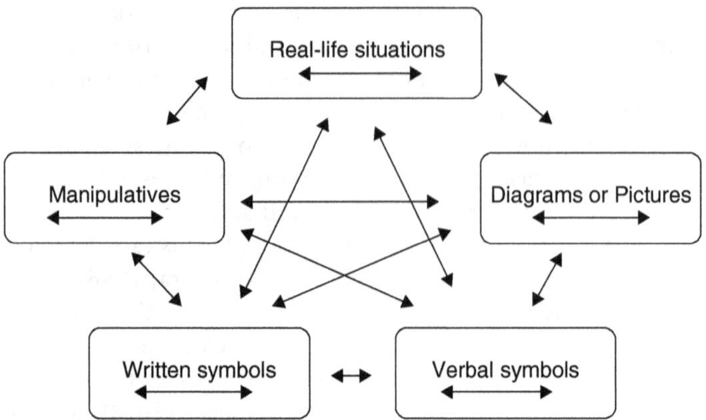

Figure 11.1 Multiple representations translations model (LMRTM) (Lesh, 1987).

students' internal conceptualisations" (Lesh et al., 1987, p. 34). The LMRTM model argues that the ability to make translations between and within various modes of representation shown in Figure 11.1 reflects the learners' understanding of mathematical concepts. This model suggests that if a learner understands a mathematical idea, she or he should have the ability to make translations between and within modes of representations.

The model describes five distinct modes of representation (see Figure 11.1) that occur when learning mathematics. These are first the real-world situations where knowledge is organised around 'real world' events. Second are manipulatives in teaching and learning in which the 'elements' in the system have little meaning, but the 'built in' relationships and operations fit many everyday situations. Third are pictures or diagrams – static or dynamic figural models. Fourth is the spoken language or verbal symbols which can be everyday language. Last are written symbols in which specialised sentences and phrases take place (Lesh et al., 1987). In the LMRTM model in Figure 11.1, the translations among modes and transformations within them are important. The translations among representations are crucial as they require learners to establish a relationship from one representational system to another. They show how interlinked the representations are to each other during the process of teaching and learning. As noted by van de Walle (2004), learners who have difficulty translating a concept from one representation to another tend to have problems in solving problems and understanding computations. Thus, teachers need to work towards strengthening learners' abilities to navigate between and among these representations in their efforts to help learners improve their conceptual understanding in mathematics.

As research suggests, this model is important because it shows that between mathematics lessons teachers should adjust the teaching models they use and give cognisance to the multiple ways of learning that each learner has (Nugroho et al., 2019). This is because each learner is unique and the way they learn differs.

Thus, providing a multimodal system during instruction will benefit the majority of learners, especially in a multilingual class. Many studies have also highlighted that multiple-representations-based mathematics instruction enhances understanding of abstract mathematical concepts (Ainsworth, 2006), provides meaningful mathematical understanding (Dreher & Kuntze, 2015) and contributes to learners' conceptual construction of mathematical knowledge (Goldin, 2004; NCTM, 2008).

Research Methodology

The data for this study comprise diagrams, verbal language and a set of gestures displayed by a selected teacher Mr Qanda (not his real name), while teaching trigonometry to a multilingual Grade 11 mathematics class. Mr Qanda is a qualified high school mathematics teacher who is an isiXhosa first language speaker. I videotaped him teaching five trigonometry lessons to a Grade 11 class in the Eastern Cape province of South Africa. At the end of the five lessons, I interviewed Mr Qanda once for about an hour. I worked and read through video recordings and transcriptions and noted the language, diagrams and the nature of the gestures that he demonstrated, in addition to their similarities and differences. The diagrams, verbal language and gestures used by Mr Qanda were isolated and identified. An analytical framework, using McNeill's (1992) classification of gestures (iconic, pointing, metaphoric and beat), was used to categorise his gestures.

Findings

The use of diagrams, gestures and verbal language in both the LOLT and the learners' first language was evident during the teaching of trigonometry in Grade 11. Diagrams were mostly drawn on the chalkboard and were used to explain trigonometric concepts. IsiXhosa was used together with the LOLT (English), through code switching during teaching. Mr Qanda used mainly iconic and pointing gestures in conjunction with the learners' first language (isiXhosa) and the LOLT.

In the interview, Mr Qanda mentioned:

> ... I usually draw diagrams on the chalkboard, sometimes I bring charts with drawn diagrams. These I use because they help simplify explanations. During teaching I also use their first language although I try to use English most of the time.

Mr Qanda also said:

> ... during lessons 3 and 5 as you have seen, I had to trace the trigonometric graphs in the air with my finger to show them the basic outline of the cosine graph and the tan graph.

The concurrent use of diagrams, verbal and gestural expressions was evident in this teacher's lessons in the multilingual classes (see Excerpts 11.1 and 11.2). The advantage of this was that it allowed Mr Qanda to show his visualisation of concepts through diagrams, gestures and other representations of trigonometry concepts during teaching. Thus, the concurrent use of diagrams, various gestures and verbal communication in both isiXhosa and the LOLT was seen to benefit the teaching of trigonometry in these multilingual classes.

In Excerpt 11.1, Mr Qanda's lesson focused on developing the formulae for the area rule. Mr Qanda illustrated the included angle and its related sides using his hands (iconic-symbolic gesture) during the discussion of the area rule. Hand gestures such as pointing, shape, movement and tracing were used by Mr Qanda to provide a representation of an included angle. He used isiXhosa during this process when he asked "*Jonga, zeziphi i-sides endisinidayo? Ngu-c nabani?*" (Look, which sides do I need to consider? It's *c* and what?). When giving this explanation, hand movements were used as they formed the shape of an included angle (see Excerpt 11.1). This interplay of representations was common in his lessons and deemed necessary to engage learners in understanding the area rule informally before stating it as a formula. Throughout his lesson, both isiXhosa and English (the LOLT) were used simultaneously.

Excerpt 11.1: Mr Qanda's Gestures and Verbal Language (Lesson 2)

Visuals	*Description*
Gestures Pointing Iconic-symbolic (shape and movement) Tracing **Diagrams** Triangles were used to prove the area rule	Mr Qanda used his hands to point to the aspect under discussion, to make lines and angles in the air imitating the angle he was referring to on the diagram. He did this while speaking in isiXhosa. His hands also moved to show referent sides of the diagram.

Mr Qanda's words during teaching

Nantsi i-clue ukuba ndi sebenza ngo-angle (Here is a clue when working with angle A). [Pointing at angle *A* , see (a).]

Jonga, zeziphi i-sides endizinidayo? Ngu-c nabani? No-b. (Look, which sides do I need to consider? Its *c* and what? And *b*.) [Teacher moves both his hands along *AB* and *BC* ; see (b) and (c).]

Uba ndisebenza ngo- angle C, ngubani endimnidayo? Ngu-a, nabani? No-b. Ubandisebenza ngo-angle B, ngubani endimnidayo? (If I use angle C, which sides do I need to consider? It is '*a*' and which side? And '*b*' . If I use angle *B* , which sides do I need to consider)? [Teacher moves both his hands along *AB* and *BC* (see e) and f).] [chorus answer from the learners] ***a....a*** [teacher points at a learner. The learner responds ***A and C***.]

Excerpt 11.2: Mr Qanda's Gestures and Verbal Language (Lesson 1)

Visuals	*Description*
Gestures Pointing (action; spatial position) Iconic-symbolic (spatial position; shape) **Diagrams** Different types of triangles	Mr Qanda used figures to show the diagram of an included angle. Then he showed the two sides and the angle using pointing gestures. He pointed at height *h* which he had just drawn. He showed what he wanted to emphasise. Action gestures were then used in horizontal movements of his hands, to show the base of the triangle. He used his hands to show what would be regarded as the base of this triangle since any side may be regarded as the base.

Mr Qanda's words during teaching

Nanku mzekelo i-condition yokuqala ndine right angle triangle, nantsi triangle yam. (Here is an example, the first condition is that I have a right angle[d] triangle, here is my triangle.) [The teacher draws the triangle.]

Ndinikwe eli side neli side nale- angle and aku specifayiwanga, akuthwanga phaya – an included angle, kuthwe two sides and an angle. (I'm given this side and this side, and this angle it has not been specified it is and included angle, it just says two sides and an angle.)

For i-area rule, masikhangele Sithe ukubangaba ndino triangle ABC ndibe ne le-height, mhlambe le height ndiyiphi gama, ndithi ngu 'h' neh. (For the area rule, let's look we said if I have triangle ABC and I also have this height let's say I give this height a name and call it '*h*' right)? [In in this teacher was pointing to the shape on the chalkboard and the specific sections of reference.]

Kule triangle ABC, I know ukubangaba lona ngu angle C, this side will be side c. Ubangaba this is B, icala lonke leli ngu 'b'. Yi base ke mos? So, i base ayingo 'd', but ngu- 'b'. (In this triangle ABC I know that if this is angle C, this is side c this is the base. If this is B, this whole side is b. That is the base right? So, the base is not D, it is b, using pointing and illustrating gestures.)

Mr Qanda code switched frequently in his lessons as can be seen in Excerpts 11.1, 11.2 and 11.3. This was done to help learners follow and understand trigonometric concepts. An example from Excerpt 11.3 is when he said "*Apha ngoku asisabhali 'h' si za bhala bani, ngubani u 'h' wethu ngoku? Ngu c.Sin A*" (In this slot we no longer write '*h*'. What do we write, what is our '*h*' now? It is c.SinA.) Here *h*'s equivalent was found by changing the subject of the formula. Mr Qanda used isiXhosa at the same time as he referred to the diagram on the chalkboard (see Excerpt 11.3). Mr Qanda simultaneously used his fingers and diagrams (see Excerpt 11.2). Iconic representation was evidenced in this case by the teacher and I consider these lessons as examples of multiple modes of representation (see Excerpts 11.1–11.3) as proposed by Lesh et al. (1987). Mr Qanda used isiXhosa, the LOLT (English), gestures and diagrams as shown in

Excerpt 11.3: Mr Qanda's Gestures and Verbal Language (Lesson 3)

Visuals	*Description*
Gestures Pointing Tracing Iconic-symbolic (shape and movement) **Diagrams** Triangles	Uphi u a wethu? He used his hands to show the side of the triangle he was referring to. He did the same with the written expression on the second diagram. He used hand gestures to clear some misconceptions that could arise. Here Mr Qanda linked the variables a and b, and the angle C in the formula to the diagram. He showed and emphasised the three trigonometrical ratios. He used his hands to give options from which learners could choose – sin , cos .

Mr Qanda's words during teaching

Isiphinto ethi kengoku uba sifunu 'h' as the subject of the formula. Jonga, siza cross multiplier akunjalo? Ekuqhibeleni u 'h' will be H= c.* sin *A. Can you see now? That represents kanye le value ka 'h'. Apha ngoku asisabhali 'h' si za bhala bani, ngubani u 'h' wethu ngoku? Ngu c.*Sin *A. (This gives us the direction that if we are looking for h as the subject of formula. Look we will cross-multiply isn't it? In the end $h = c.\sin A$. That represents precisely the value of h . In this slot we no longer write "h" what do we write, what is our "h" now? It is $c.\sin A$.)

Masikhangele, uphu : 'a' wethu? Nanku! U side 'a' lona. Uphu side, that is side 'a' the second one. Remember there was 'b' which is the normal base is 'b'. But u 'h' wethu ngu a.SinC. (Let us look, where is our 'a' ? Here is it. This is side 'a' the second one. But our h is $a.\sin C$.)

Uphi u 'a' lo 'b'? There is 'a' and 'b' and nantsi angle yethu. Now, sisebenze ngo angle A, sa sebenza ngo angle C. What will be the correct formula for, if you use angle B? Khakucinge, sisebenza ngo angle bani? Ngo angle B $\frac{1}{2}$***. a. c.* Sin*B*** . (Where is 'a' and 'b :"? There is a and b and here is our angle. Now we have worked with angles A and C . Think about it, which angle are we dealing with? Its B hence $\frac{1}{2}. a. c. \text{Sin}B$.)

the excerpts. This was also the case for all the observed lessons in this study. The drawing of diagrams on the chalkboard, the use of various gestures (pointing, shape, movement, tracing and iconic) and the teacher's use of verbal language in both isiXhosa and English was observed as a frequent occurrence in Mr Qanda's lessons. IsiXhosa words such as *jonga* (look), *siyabona sonke* (can we all see), *siyayibona kwi-diagram* (can we see on this diagram) *masikhangela lapha* (when we look here) (see Excerpts 11.1–11.3), among others, were used to draw the learners' attention to the part of the diagram or writing on the chalkboard that they were focusing on, together with deictic gestures. I believe that the teacher's actions were mainly meant to ground the trigonometric concepts in the classroom environment so that the learners could focus on the crucial aspects of the lesson. I argue that all his actions were Mr Qanda's endeavours to help his learners access trigonometric concepts in this class.

Conclusion

In this chapter, the secondary school teacher I focused on predominantly used pointing and iconic (shape, movement, tracing) gestures, diagrams drawn on the chalkboard and verbal language in isiXhosa and English concurrently, while teaching. The use of such multiple representations was apparently used to ground teaching and learning in the physical environment of the classroom (Clough & Hilverman, 2018). This prevalence of pointing and iconic gestures, used simultaneously with diagrams, could be what Valenzeno et al. (2003) referred to: that "pointing and tracing gestures 'ground' teachers' speech by linking abstract, verbal utterances to the concrete, physical environment" (p. 187). While other forms of gestures where used, they were not as prevalent. Other forms of gestures were used by other teachers to direct and focus attention to the features of the problem, through constant reference to the drawn diagrams. The teachers also used gestures to draw lines and angles in the air (tracing) in addition to draw attention to relevant information during teaching.

The two-directional and reciprocal relation that occurred through, for example, iconic-representational gestures during the teacher's demonstration of an included angle was vital, especially considering the multilingual nature of the class he was teaching. This may be considered as a semiotic tool that was used to help learners to perceptually encode the idea of 'inclusion' because of the two adjacent sides. I argue that this was also the teacher's way of providing an inclusive learning environment where no learner was to be left behind. The pointing gestures directed at specific angles, and movement and tracing gestures along the sides of reference or other components on drawn diagrams during the teaching of trigonometry may be considered as a way of guiding learners' perceptual encoding processes. I also argue that teachers' synchronous use of visualisation processes (such as gestures, diagrams and verbal language in their learners' mother tongue) during teaching and learning mathematics in rural and township schools can help bridge the inequality gap between schools whose learners' LOLT is not well developed and those that are not in the Southern African region.

I conclude that in multilingual classes the interplay of all forms of teacher visualisation processes such as gestures, objects drawn on the chalkboard, charts as diagrams, or actual objects and their locations in reality, together with the correct use of verbal language in both the LOLT and mother tongue, plays a major role in increasing epistemological access to mathematical concepts at high school level. This was also concluded by Goldin-Meadow et al. (2009) that gestures referring to diagrams or objects in physical referents in monolingual settings make it easier for teachers to help learners link their developing mental representations to relevant parts of the external environment. Diagrams, gestures and language (first language or LOLT) complement each other and should be used as resources during teaching. Various representations need to be thought of as a single system or bundle, larger than either diagrams or language or gesture when considered separately (McNeill, 1992), especially during the teaching of multilingual classes. The multilingual mathematics classroom should be as flexible as possible in terms of language and various visual tools should be used when teaching, to enable and enhance epistemological access to mathematics concepts.

References

Adler, J. (2001). *Teaching Mathematics in Multilingual Classrooms.* Dortrecht: Kluwer Academic Publisher.

Ainsworth, S. (2006). DeFT: A conceptual framework for considering learning with multiple representations. *Learning and Instruction*, 183–198. http://dx.doi.org/10.1016/j.learninstruc.2006.03.001

Alibali, M. W., Nathan, M. J., Church, R. B., Wolfgram, M. S., Kim, S., & Knuth, E. J. (2013). Teachers' gestures and speech in mathematics lessons: Forging common ground by resolving trouble spots. *ZDM Mathematics Education*, *45*, 425–440. https://doi.org/10.1007/s11858-012-0476-0

Arcavi, A. (2003). The role of visual representations in the learning of mathematics. *Educational Studies in Mathematics, 52*(3), 215–241. https://doi.org/10.1023/A:1024312321077.

Arzarello, F. (2006). Semiosis as a multimodal process. *RELIME. Revista latinoamericana de investigación en matemática educativa*, *9*(1), 267–300.

Arzarello, F., & Robutti, O. (2004). Approaching functions through motion experiments. *Educational Studies in Mathematics 57*(3), 305–308.

Arzarello, F., Paola, D., Robutti, O., & Sabena, C. (2009). Gestures as semiotic resources in the mathematics classroom. *Educational Studies in Mathematics, 70*(2), 97–109. https://www.jstor.org/stable/40284563

Bender, J., & Marrinan, M. (2010). *The Culture of Diagram.* Stanford: Stanford University Press.

Carter, J. (2018). The role of diagrams in contemporary mathematics: Tools for discovery? In P. Chapman, G. Stapleton, A. Moktefi, S. Perez-Kriz, & F. Bellucci (Eds.), *Diagrammatic Representation and Inference: 10th International Conference, Edinburgh, UK, June 18–22, 2018* (pp. 787–790). Cham: Springer.

Chikiwa, C., & Schäfer, M. (2019). Visualisation processes in mathematics classrooms - the case of gestures. In M. Graven, H. Venkat, A. A. Essien, & P. Vale (Eds.), *43rd Annual Meeting of the International Group for the Psychology of Mathematics Education. 4* (pp. 4–24). Pretoria, South Africa: PME.

Cienki, A., & Müller, C. (2008). *Metaphor and Gesture.* Amsterdam: John Benjamins Publishing.

Clough, S., & Hilverman, C. (2018). Hand gestures and how they help children learn. *Front. Young Minds, 6*(29). https://kids.frontiersin.org/articles/10.3389/frym.2018.00029

Cochet, H., & Vauclair, J. (2014). Deictic gestures and symbolic gestures produced by adults in an experimental context: Hand shapes and hand preferences. *Laterality, 19*(3), 278–301. 10.1080/1357650X.2013.804079.halshs-01464412

Cohen, C. A., & Hegarty, M. (2007). Individual differences in use of external visualisations to perform an internal visualisation task. *Applied Cognitive Psychology, 21*(6), 701–711. http://dx.doi.org/10.1002/acp.1344

Diezmann, C., & English, L. (2001). Promoting the use of diagrams as tools thinking. In A. A. Cuoco (Ed.), *The roles of representation in school mathematics* (pp. 77–89). Reston, VA: National Council of Teachers of Mathematics.

Dreher, A., & Kuntze, S. (2015). Teachers facing the dilemma of multiple representations being aid and obstacle for learning: evaluations of tasks and theme-specific. *Journal für Mathematik-Didaktik, 36*(1), 23–44. http://dx.doi.org/10.1007/s13138-014-0068-3

Edwards, L. D. (2009). Gestures and conceptual integration in mathematical talk. *Educational Studies in Mathematics, 70*, 127–141. https://www.jstor.org/stable/40284565

Ekman, P., & Friesen, W. V. (1974). Detecting deception from the body or face. *Journal of Personality and Social Psychology, 29*(3), 288–298. https://doi.org/10.1037/h0036006

Gibson, J. J. (2014). *The Ecological Approach to Visual Perception: Classic Edition.* New York: Taylor & Francis.

Goldin-Meadow, S. (2004). Gesture's role in the learning process. *Theory Into Practice, 43*, 314–321. https://muse.jhu.edu/article/175836

Goldin-Meadow, S. (2015). From action to abstraction: Gesture as a mechanism of change. *Developmental Review: Dr, 38*, 167–184. https://doi.org/10.1016%2Fj.dr.2015.07.007

Goldin-Meadow, S., & Alibali, M. W. (2013). Gesture's role in speaking, learning, and creating language. *Annual Review of Psychology, 64*, 257–83. https://doi.org/10.1146/annurev-psych-113011-143802

Goldin-Meadow, S., Cook, S. W., & Mitchell, Z. A. (2009). Gesturing gives children new ideas about math. *Psychological Science, 20*(3), 267–272. https://doi.org/10.1111%2Fj.1467-9280.2009.02297.x

Gutiérrez, A. (1996). Visualization in 3-dimensional geometry: In search of a framework. In L. Puig, & A. Gutierrez (Eds.), *Proceedings of the 20th International Conference for the Psychology of Mathematics Education, Vol. 1* (pp. 3–19). Valencia, Spain.

Gutiérrez, R. (2018). Why we need to rehumanize mathematics. In I. Goffney, R. Gutiérrez, & M. Boston (Eds.), *Annual Perspectives in Mathematics Education: Rehumanizing Mathematics for Students Who Are Black, Latinx, and Indigenous.* National Council of Teachers of Mathematics. Reston, Virginia.

Hare, A., & Sinclair, N. (2015). Pointing in an undergraduate abstract algebra lecture: Interface between speaking and writing. In K. Beswick, T. Muir, T., & J. Wells (Eds.), *Proceedings of PME Hobart, Australia, 39, 3*, 33–40.

Kadunz, G. (2006). Experiments with diagrams—a semiotic approach. *Zentralblatt für Didaktik der Mathematik, 38*, 445–455. https://doi.org/10.1007/BF02652781

Kendon, A. (2004). *Gesture: Visible Action as Utterance.* Cambridge: Cambridge University Press.

Koedinger, K. R., Alibali, M. W., & Nathan, M. J. (2008). Trade-offs between grounded and abstract representations: Evidence from algebra problem solving. *Cognitive Science, 32*(2), 366–397. https://doi.org/10.1080/03640210701863933

Kress, G. (2004). Reading images: Multimodality, representation and new media. *Information Design Journal, 12*(2), 110–119. https://doi.org/10.1075/idjdd.12.2.03kre

Lesh, R., Post, T., & Behr, M. (1987). Representation and translations amongst representations in mathematical learning. In C. Janvier (Ed.), *Problems of Representation in the Teaching and Learning of Mathematics* (pp. 33–40). Hillsdale: Lawrence Earlbaum.

Lim, F. V. (2019). Analysing the teachers' use of gestures in the classroom: A Systemic Functional Multimodal Discourse Analysis approach. *Social Semiotics, 29*(1), 83–111. https://content.apa.org/doi/10.1037/h0036006

McNeill, D. (1992). *Hand and Mind: What Gestures Reveal About Thought.* Chicago, IL: University of Chicago Press.

McNeill, D. (2005). *Gesture and Thought.* Chicago: University of Chicago Press.

Mesquita, A. (1998). On conceptual obstacles linked with external representation in geometry. *Journal of Mathematical Behavior, 17*(2), 183–195. https://doi.org/10.1016/S0364-0213(99)80058-5

Morrow, W. (2007). *Learning to Teach in South Africa.* Cape Town: HSRC Press.

Mudaly, V. (2012). Diagrams in mathematics: To draw or not to draw? *Perspectives in Education, 30*(2), 22–31. https://journals.ufs.ac.za/index.php/pie/article/view/1756

Nardi, E. (2014). Reflections on visualisation in mathematics and in mathematics education. In M. N. Fried, & T. Dreyfus (Eds.), *Mathematics & Mathematics Education: Searching for Common Ground. Advances in Mathematics Education* (pp. 193–220). Dordrecht: Springer.

National Council of Teachers of Mathematics [NCTM]. (2008). *The Role of Technology in the Teaching and Learning of Mathematics.* Reston, VA: NationalCouncil of Teacher of Mathematics.

Nugroho, M. A., Budiyono, & Slamet, I. (2019). Experimentation of innovative learning models in terms of students multiple ıntelligences among middle school students in Demak District. *International Journal of Educational Research Review, 4*(2), 262–268.

Nunokawa, K. (2006). Using drawings and generating information in mathematical problem solving processes. *Eurasia Journal of Mathematics, Science and Technology Education*, *2*(3), 34–54. www.ejmste.com

Pantziara, M., Gagatsis, A., & Elia, I. (2009). Using diagrams as tools for the solution of non-routine mathematical problems. *Educational Studies in Mathematics*, *72*(1), 39–60. http://dx.doi.org/10.1007/s10649-009-9181-5

Radford, A., Atkinson, M., Britain, D. J., Clahsen, H., & Spencer, A. (2009). *Linguistics: An Introduction.* Cambridge: Cambridge University Press.

Sabena, C. (2018). Exploring the contribution of gestures to mathematical argumentation processes from a semiotic perspective. In G. Kaiser, H. Forgasz, M. Graven, A. Kuzniak, E. Simmt, & B. Xu (Eds.), *Invited Lectures from the 13th International Congress on Mathematical Education* (pp. 541–559). Cham: Springer.

Sáenz-Ludlow, A., & Kadunz, G. (Eds.). (2016). *Semiotics as a Tool for Learning Mathematics: How to Describe the Construction, Visualisation, and Communication of Mathematical Concepts.* Rotterdam, The Netherlands: Sense Publishers.

Scherer, K. R., & Ekman, P. (Eds.). (1982). *Handbook of Methods in Nonverbal Behavior Research.* Cambridge: Cambridge University Press.

Sinclair, N., & Tabaghi, G. S. (2010). Drawing space: Mathematicians' kinetic conceptions of eigenvectors. *Educational Studies in Mathematics*, *74*(3), 223–240. https://www.jstor.org/stable/27822715

Tall, D. O. (2004). The three worlds of mathematics. *For the Learning of Mathematics, 23*(3), 29–33.

Valenzeno, L., Alibali, M. W., & Klatzky, R. (2003). Teachers' gestures facilitate students' learning: A lesson in symmetry. *Contemporary Educational Psychology, 28*, 187–204. https://psycnet.apa.org/doi/10.1016/S0361-476X(02)00007-3

Van de Walle, J. (2004). *Elementary and Middle School Mathematics: Teaching Developmentally, 5th Edition.* New York: Pearson.

Winn, W. (1987). Charts, graphs, and diagrams in educational materials. In D. M. Willows, & H. A. Houghton (Eds.), *The Psychology of Illustration: Volume 1. Basic Research.* New York: Springer.

Yerushalmy, M., & Naftaliev, E. (2011). Design of interactive diagrams structured upon generic animations. *Technology, Knowledge and Learning, 16*(3), 221–245. https://doi.org/10.1007/s10758-011-9183-0

Yoon, C., Thomas, M., & Dreyfus, T. (2011). Grounded blends and mathematical gesture spaces: Developing mathematical understandings via gestures. *Educational Studies in Mathematics, 78*, 371–393. https://doi.org/10.1007/s10649-011-9329-y

12 Final Word – A Synthesis

Marc Schäfer

In each chapter, the individual authors interrogated how visualisation processes can be incorporated in the teaching and learning of mathematics in unique classroom settings and contexts. These were mainly in South Africa and Namibia, and also included experiences in Italy and Germany. In all the chapters the authors attempted to explore how the harnessing of visualisation processes could enhance epistemological access to mathematics, which to many learners remains elusive. Each chapter was framed by a particular case study that tells a story of how mathematics teachers and learners interacted with each other through different visualisation media in their attempt to make sense of mathematical ideas and concepts. Collectively, the individual stories and vignettes formed a bigger narrative of how the mathematics classroom can be an interesting and exciting space for learning and exploration for everyone, no matter in what socio-political context the classroom is situated.

The book began with a brief contribution towards finding possible visualisation theoretical framings that could inform visualisation research agendas. Apart from exploring possible theoretical underpinnings, Mudaly and Schäfer argued that adopting a strong visualisation agenda in the mathematics classroom enables a strong and meaningful mathematical mediation. Each chapter was thus undergirded by a strong and explicit visualisation agenda.

Part 1: Visualisation and Pedagogy

In his chapter on teaching number sense, Griqua's research findings showed that adopting an enhanced visual teaching approach to number sense privileges conceptual understanding over procedural fluency. He thus argued that, in the context of his study, once conceptual understanding is achieved in the number sense, learners better understand why the accompanying procedures work, and are then able to deduce the necessary rules by themselves.

In her chapter on teaching fractions, Ausiku and her group of primary school teachers experimented with different models of fractions in their mediation of fraction concepts that learners find difficult to understand. Despite the dominance of the area model, the length and number line models were also prevalent in some of the participants' lessons. In most instances, they relied on drawing

DOI: 10.4324/9781003172420-16

their own circles, rectangles, number lines, sets of objects and concrete objects to explain fraction concepts. The use of chalkboard visuals dominated their practice. Findings suggested that, in the context of this case study, fraction visuals were inconsistently used and that the area model was the dominant fraction model employed. Also, there was an imbalance between the verbal and the non-verbal (visual) codes. In their reflections the participants agreed that the incorporation of more visuals improves their pedagogy. They also suggested that the alignment of visuals with the lesson objectives is important, as is the need to make explicit the link between the non-verbal and the verbal codes.

Part 2: Visualisation and Learning

Findings in Samson's case study – which explored how selected learners were able to generalise pictorial patterns in multiple ways using a variety of visually mediated approaches – suggested that semiotic means such as gestures, rhythm and speech are not simply outward manifestations of thinking and internal workings, but integral processes in the formation of knowledge. Micro-analyses of learners' whole-body engagement with visual patterning tasks revealed not only generalisation approaches that evolved out of a purposeful and conscious search for structure, but also strategies and visual apprehensions that emerged and developed serendipitously from unstructured exploration and interaction with the pictorial context. Samson argued that there is a strong moral or ethical dimension to embracing a multi-semiotic notion of embodied understanding. This ethical dimension points to the idea that different pupils have different learning styles, different ways of engaging with or making sense of mathematical situations and different ways of *seeing* the world. The recognition of this diversity is key in addressing learning equity and ultimately epistemological access.

The relationship between visualisation and reasoning seems obvious – without visualising or imagining, there can be no reasoning and vice versa. Dongwi's chapter interrogated this notion in some depth in the context of a group of learners collaboratively solving a given word problem. The evidence of the study suggested that visualisation and reasoning are indeed inseparable, that is, they are intertwined in such a way that they in fact co-emerge. Dongwi showed the inseparability between these two processes by carefully analysing the interactions and argumentations between her participants as they solved their given problems. She argued that teachers should thus take into account both verbal and non-verbal (visual) thinking processes of their learners as they engage in problem-solving activities.

Herbert also argued for a strong emphasis in visual activities for the conceptual understanding of algebra. In her study, which took place in the context of an after-school club, her participants were required to create and engage in visual representations of linear algebraic expressions in their attempt to understand and comprehend the meaning of these expressions. The multitude of visual representations that her participants devised were, in part, remarkably innovative and original.

Part 3: Visualisation and Technology

The cluster of case studies that involved Information and Computer Technology (ICT) all showed that the strategic harnessing of the inherent visual nature and properties of digital appliances can be utilised as very effective learning and teaching devices and media in a mathematics classroom. Through computer screen tracking software, Mavani, in his study, tracked the workings of learners as they explored and came to grips with selected Euclidean geometry theorems. He showed how learners, who participated in an ICT intervention programme, navigated through the computer screen and utilised the dynamic capacity of *GeoGebra* to test their conjectures and results in ultimately arriving at suitable and appropriate solutions.

As mobile technology devices such as smartphones are becoming more compact, ubiquitous and integral to our daily routines and activities, so are they becoming more useful in educational settings, as tools for both mediation and learning. The use of smartphones has opened up many cost-effective platforms for accessing and disseminating information on a global scale, thereby creating opportunities for epistemological access, even in the remotest areas of the globe with minimal access to traditional resources such as books and sundry educational learning and teaching kits. Chikiwa and Ludwig's chapter documented aspects of work done in the context of the MathsCityMap project which utilises the smartphone as a tool to navigate outdoor mathematics trails and mediate mathematical problem-solving experiences in original places strategically situated *en route* of the mathematics trail. Chikiwa and Ludwig provided evidence that participating in outdoor mathematics trails has the potential to enhance learning and promote improved access to mathematical concepts. The outdoor tasks that are at the heart of the trails have advantages of situatedness and hence they become meaningful, visual, stimulating, challenging, and exciting for learners. This authentification of mathematical problems provided an inclusive learning environment – thus affording meaningful potential to enhance epistemological access for all learners. Chikiwa and Ludwig asserted that outdoor tasks through mathematics trails invited all learners, irrespective of their classroom achievement level, to participate successfully in problem-solving activities and gain a sense of pride in the mathematics they create.

Part 4: Visualisation, Semiotics and Language

Sabena and Schäfer's work makes a strong case for the recognition and acknowledgement that the process of teaching and learning is an embodied one. They provided evidence that bodily actions – in particular gestures – are integral to the entire cognitive and pedagogical process.

They argued that gestures are integral to the semiotic bundle and may contribute, in synergy with other semiotic resources, to mathematics reasoning and communication. They proposed that these bodily actions are manifestations of the entire body's role in generating and understanding knowledge, and should thus prompt teachers to overcome the hegemony of speech and symbols, and broaden their focus of attention in the classroom, so as to also include students' gestures as part and parcel of their mathematical activity. In parallel, Chikiwa's case study showed how language and gestures are interwoven and form a multimodal system of signs and symbols in the teaching of mathematics. This multimodality has particular relevance and significance in multilingual classrooms where teachers' synchronous use of visualisation processes such as gestures, diagrams and verbal language (in their learners' mother tongue) can help bridge the inequality gap between those learners whose language of learning and teaching is well developed and those whose is not. Chikiwa argued that in multilingual classes the interplay of all forms of teacher visualisation processes such as gestures, objects drawn on chalkboard or charts as diagrams, or actual objects and their locations in reality, together with the appropriate use of verbal language in both the official language of learning and teaching and the mother tongue, plays a major role in increasing epistemological access to mathematical concepts at high school level.

As a final word, the question needs to be asked whether the original objective of the book, namely, to deliberate and examine how the use of visualisation processes in the mathematics classroom can enhance both teaching and learning by drawing on classroom research conducted in the Southern African region, has been addressed. Each chapter provided some insight into a number of classroom situations across the spectrum of schools in Southern Africa, in an attempt to interrogate how epistemological access in a context of gross inequality can constructively be addressed by providing research-based solutions and recommendations on how adopting a visualisation agenda can play a part in facilitating this access.

I wish to acknowledge the significant financial contribution of the South African National Research Foundation (NRF) in supporting the research activities of many of the researchers in this book. Their individual stories will hopefully inspire other researchers to contribute to a shared visualisation narrative that has the potential to contribute to the transformation of the mathematics classroom in this region.

The opinions and arguments of the authors are not necessarily those of the NRF.

Index

Note: **Bold** page numbers refer to tables, *italic* page numbers refer to figures.

Taylor & Francis eBooks

www.taylorfrancis.com

A single destination for eBooks from Taylor & Francis with increased functionality and an improved user experience to meet the needs of our customers.

90,000+ eBooks of award-winning academic content in Humanities, Social Science, Science, Technology, Engineering, and Medical written by a global network of editors and authors.

TAYLOR & FRANCIS EBOOKS OFFERS:

A streamlined experience for our library customers

A single point of discovery for all of our eBook content

Improved search and discovery of content at both book and chapter level

REQUEST A FREE TRIAL

support@taylorfrancis.com

For Product Safety Concerns and Information please contact our EU representative GPSR@taylorandfrancis.com
Taylor & Francis Verlag GmbH, Kaufingerstraße 24, 80331 München, Germany

www.ingramcontent.com/pod-product-compliance
Lightning Source LLC
Chambersburg PA
CBHW060256220326
41598CB00027B/4129